PASSING THE NORTH CAROLINA ALGEBRA I END-OF-COURSE TEST

Developed to the new 2003 Standard Course of Study!

ERICA DAY
ALAN FUQUA
COLLEEN PINTOZZI

AMERICAN BOOK COMPANY
P. O. BOX 2638
WOODSTOCK, GEORGIA 30188-1383
TOLL FREE 1 (888) 264-5877 PHONE (770) 928-2834
FAX (770) 928-7483
WEB SITE: www.americanbookcompany.com

Acknowledgements

In preparing this book, we would like to acknowledge Marsha Torrens, Mary Stoddard, and Eric Field for their contributions in editing and developing graphics for this book. We would also like to thank our many students whose needs and questions inspired us to write this text.

by American Book Company
P.O. Box 2638
Woodstock, GA 30188-1383

Printed in the United States of America
08/05 04/07

Contents

Preface

Passing the North Carolina Algebra I End-of-Course Test will help you review and learn important concepts and skills related to Algebra I. First, take the Diagnostic Test beginning on page 1 of the book. Next, complete the evaluation chart with your instructor in order to help you identify the chapters which require your careful attention. When you have finished your review of all of the material your teacher assigns, take the progress tests to evaluate your understanding of the material presented in this book. **The materials in this book are based on the 2003 standards that are published by the North Carolina Department of Education. The complete list of standards is located in the Answer Key. Each question in the Diagnostic and Practice Tests is referenced to the standard, as is the beginning of each chapter.**

This book contains several sections. These sections are as follows: 1) A Diagnostic Test; 2) Chapters that teach the concepts and skills for ***Passing the North Carolina Algebra I End-of-Course Test***; and 3) Two Practice Tests. Answers to the tests and exercises are in a separate manual.

ABOUT THE AUTHORS

Erica Day has a Bachelor of Science Degree in Mathematics. She graduated with honors from Kennesaw State University in Kennesaw, Georgia. She is currently pursuing a Master of Science Degree in Mathematics at Georgia State University. She has also tutored all levels of mathematics, ranging from high school algebra and geometry to university-level statistics, calculus, and linear algebra. She is currently writing and editing mathematics books for American Book Company.

Alan Fuqua has a Bachelor of Chemical Engineering degree from the Georgia Institute of Technology and a Bachelor of Science Degree in Mathematics. from Kennesaw State University in Kennesaw, Georgia. He has over fifteen years of industrial experience in quality systems using statistical techniques as well as extensive teaching and tutoring experience at the middle and high school level. He has also been a speaker at math conferences all over the country, presenting teaching techniques for state specific testing programs. Alan and Erica have coauthored over ten books to help students pass graduation and end of course exams, including *Passing the Georgia Algebra I End of Course*, *Passing the South Carolina End of Course in Mathematics*, *Passing the Tennessee Gateway in Algebra I*, and *Passing the New Jersey HSPA in Mathematics*.

Colleen Pintozzi has taught mathematics at the middle school, junior high, senior high, and adult level for 22 years. She holds a B.S. degree from Wright State University in Dayton, Ohio and has done graduate work at Wright State University, Duke University, and the University of North Carolina at Chapel Hill. She is the author of many mathematics books including such best-sellers as *Basics Made Easy: Mathematics Review, Passing the New Alabama Graduation Exam in Mathematics, Passing the Louisiana LEAP 21 GEE, Passing the Indiana ISTEP+ GQE in Mathematics, Passing the Minnesota Basic Standards Test in Mathematics,* and *Passing the Nevada High School Proficiency Exam in Mathematics*.

Diagnostic Test

1. Which of the following is equivalent to 3^{-5}?

(A) -45

(B) -15

(C) $\frac{1}{243}$

(D) 3×10^{-5}

1.01a

2. Isabella is simplifying this expression:
$2(5a + 3b - c) - 5(4a - 2b - 3c)$
The expression above is equivalent to which of the following expressions?

(A) $-10a + 16b + 13c$
(B) $-10a - 4b - 4c$
(C) $30a + b + 2c$
(D) $30a - 4b - 17c$

1.01b

3. Andrea has 10 more jellybeans than her friend Chelsea, but Andrea has half as many as Rebecca. Which expression below best describes Rebecca's jelly beans?

(A) $R = 2C + 20$
(B) $R = C + 10$
(C) $R = A + \frac{1}{2}C$
(D) $R = 2A + 10$

1.01

4. A builder is constructing a fence 85 feet long. Each section of fence contains 6 rails of wood and takes up $2\frac{1}{2}$ feet. How many rails of wood will the builder need?

(A) 102 beams
(B) 204 beams
(C) 308 beams
(D) 420 beams

1.02

5. What is the x-intercept of the following linear equation?
$3x + 4y = 12$

(A) $(0, 3)$
(B) $(3, 0)$
(C) $(0, 4)$
(D) $(4, 0)$

4.01b

6. Which of the following equations is represented by the graph?

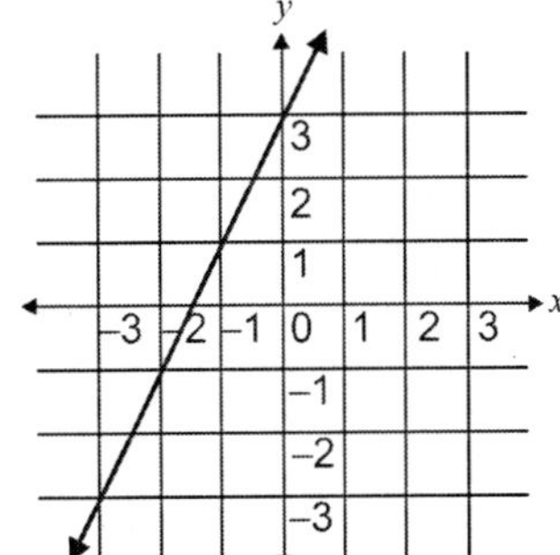

(A) $y = -3x + 3$
(B) $y = -\frac{1}{3}x + 3$
(C) $y = 2x + 3$
(D) $y = 3x - 3$

4.01a

7. Solve: $6 - 2(5y - 1) = 18$

(A) $y = 2$
(B) $y = -2$
(C) $y = 1$
(D) $y = -1$

4.01a

8. Simplify: $\sqrt{45} \times \sqrt{27}$

(A) $3\sqrt{15}$
(B) $9\sqrt{15}$
(C) $\sqrt{72}$
(D) $\sqrt{121}$

1.01a

9. Which of the following graphs represents $y = 2x^2$?

(A)

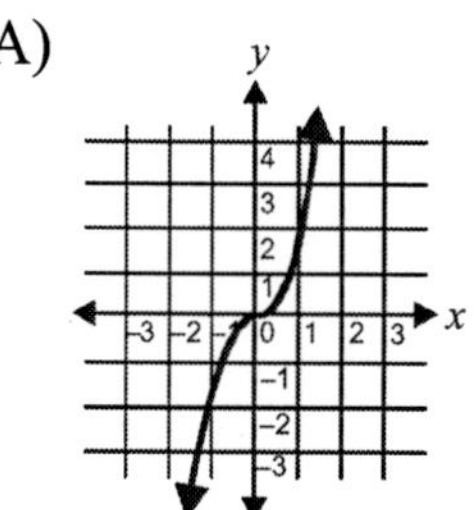

(B)

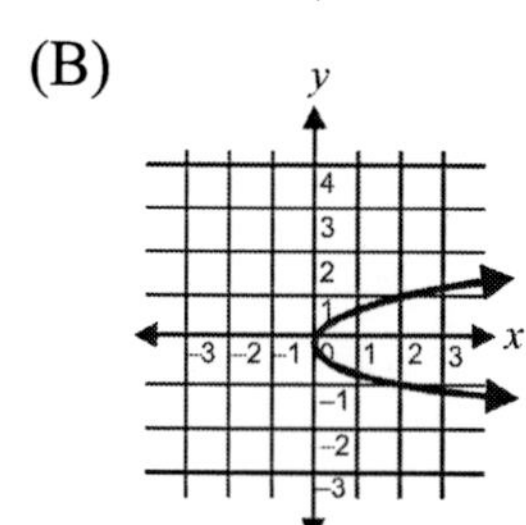

(C)

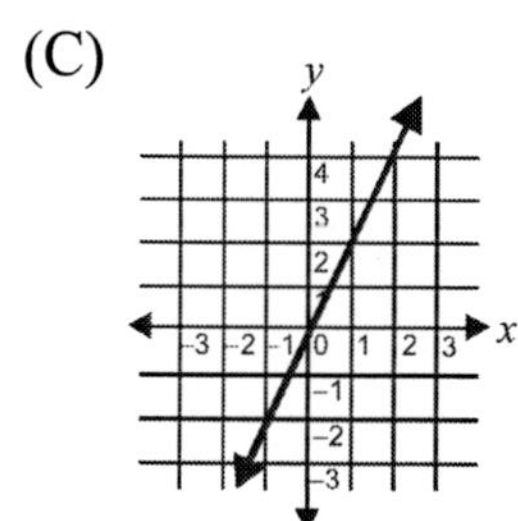

(D)

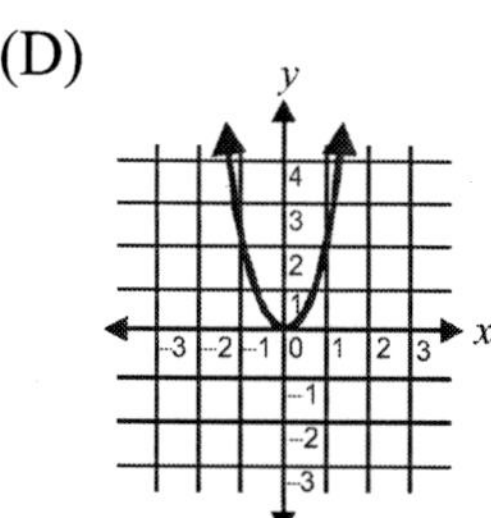

4.02

10. Solve the following inequality: $-3(4x+5) > (5x+6)+13$

(A) $x < -\frac{14}{17}$

(B) $x > -2$

(C) $x > \frac{20}{11}$

(D) $x < -2$

4.01a

11. Consider the following equations: $f(x) = 3x + 2$ and $f(x) = 3x - 7$. Which of the following statements is true concerning the graphs of these equations?

(A) The graphs of the equations are lines that are perpendicular to each other.

(B) The graph of the line represented by the equation $f(x) = 3x + 2$ always remains above the x-axis, while the graph of the line represented by the equation $f(x) = 3x - 7$ always remains below the x-axis.

(C) The graphs of the equations are lines that are parallel to each other.

(D) The graphs of the lines intersect each other at the point $(2, -7)$. 2.02

12. What is the solution to the following system of equations?
$y = 4x - 8$
$y = 2x$

(A) $(-4, -8)$
(B) $(4, 8)$
(C) $(-1, -2)$
(D) $(1, 2)$

4.03

13. Solve for a: $-2(-3-5) = 3 - a$

(A) -13
(B) 19
(C) 13
(D) -19

1.02

14. Find c: $\dfrac{c}{-2} > -6$

(A) $c > -12$
(B) $c < 12$
(C) $c > 12$
(D) $c < 3$

1.02

15. $(3x^2 - 5x + 6) - (x^2 + 4x - 7) =$

(A) $4x^2 - x - 1$
(B) $4x^2 - x + 13$
(C) $2x^2 - 9x - 1$
(D) $2x^2 - 9x + 13$

1.01b

16. In the following equation, which are the variable, the terms, and the coefficient?
$9x - 3 = 78$

(A) Variable $= x$
Terms $= 9x, -3$
Coefficient $= 9$
(B) Variable $= -3$
Terms $= 78, -3$
Coefficient $= -3$
(C) Variable $= 9$
Terms $= 9, -3$
Coefficient $= x$
(D) Variable $= 78$
Terms $= 78, 9$
Coefficient $= x$

4.01b

17. What is the value of the expression $5(x + 6)$ when $x = -3$?

(A) -9
(B) 15
(C) 9
(D) 45

1.02

18. Find the equation of the line perpendicular to the line containing the points $(-2, -3)$ and $(1, 4)$ and passing through the point $(0, 3)$.

(A) $y = \frac{3}{7}x - 3$

(B) $y = -\frac{3}{7}x + 3$

(C) $y = -\frac{3}{7}x - 3$

(D) $y = -\frac{7}{3}x + 3$

4.01a

19. An arrow shoots upward with an initial velocity of 128 feet per second. The height (h) of the arrow is a function of time (t) in seconds since the arrow left the ground and can be expressed by the equation $h = 128t - 16t^2$. When will the arrow be at a height of 240 feet?

(A) At 3 seconds and at 5 seconds
(B) Only at 3 seconds
(C) Only at 5 seconds
(D) Only at 8 seconds

4.02

20. Which table of values below represents a linear function?

(A)

x	y
0	−3
1	−1
2	1

(B)

x	y
−1	5
−2	0
1	3

(C)

x	y
0	2
−1	0
−2	−1

(D)

x	y
2	3
5	7
8	9

4.01a

21. Which equation matches the following graph?

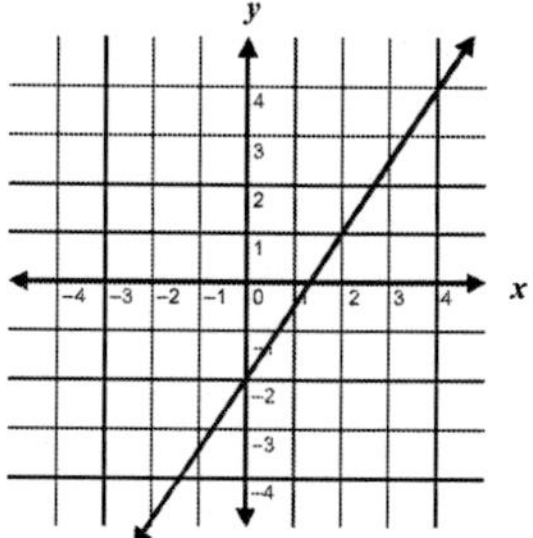

(A) $y = 3x - 2$
(B) $y + 2 = -3x$
(C) $3(y + 2) = 2x$
(D) $2(y + 2) = 3x$

4.01a

22. What graph shows a line that has a slope of -1 and a y-intercept of $(0, 1)$?

(A)
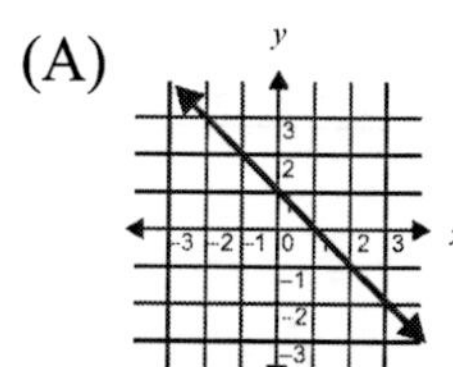

(B)
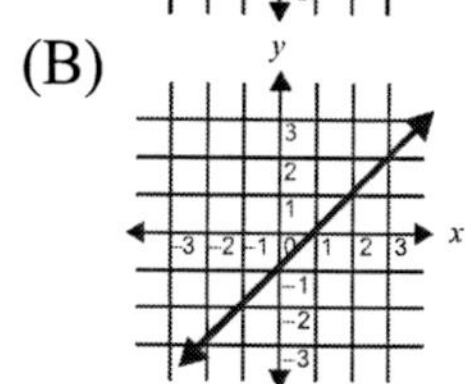

(C)
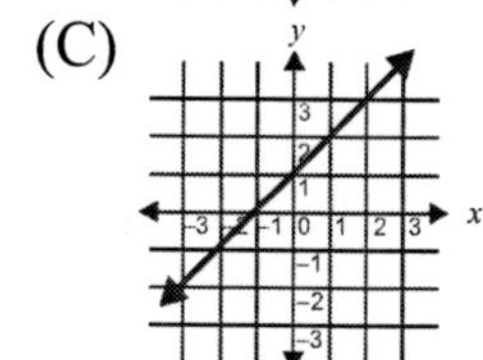

(D)
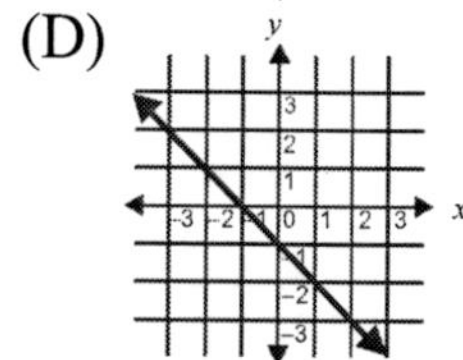

4.01b

23. Solve the equation $(x - 4)^2 = 25$.

(A) $x = -4, -5$
(B) $x = 9, -1$
(C) $x = 5, -5$
(D) $x = 4, -4$

4.02

24. Which members of the set

$$\{-3, -2, -1, 0, 1, 2, 3\}$$

are solutions for $-2x + 5 > 10$?

(A) $\{-3, -2\}$
(B) $\{0, 1, 2, 3\}$
(C) $\{-3, -2, -1\}$
(D) $\{-3\}$

4.01a

25. Solve the equation $x^2 - 6x + 7 = 0$ by completing the square.

(A) $x = 7, -1$
(B) $x = \sqrt{3}, -\sqrt{3}$
(C) $x = 3 - \sqrt{2}, \sqrt{2} + 3$
(D) $x = 3 + 2i, 3 - 2i$

4.02

26. Solve the equation $x = \sqrt{4x - 3}$.

(A) $x = 1, 3$
(B) $x = \pm\sqrt{3}$
(C) $x = -1, -3$
(D) $x = 2\sqrt{3}, -2\sqrt{3}$

4.02

27. What is the intersection of the following linear equations?

$y = 3x - 1$
$y = 4x + 2$

(A) $(-3, 10)$
(B) $(-3, -10)$
(C) $(10, -3)$
(D) $(3, -10)$

4.03

28. Evaluate $4x^3 - 2x^2 - 3x + 1$ given $x = -2$.

(A) 533
(B) 489
(C) -33
(D) 22

1.02

29. Fido gets 2 doggy treats every time he sits and 4 doggy treats when he rolls over on command. Throughout the week, he has sat 6 times as often as he has rolled over. In total, he has earned 80 doggy treats. How many times has Fido sat?

(A) 4
(B) 5
(C) 24
(D) 30

4.03

30. Which equation does the graph of the line below represent?

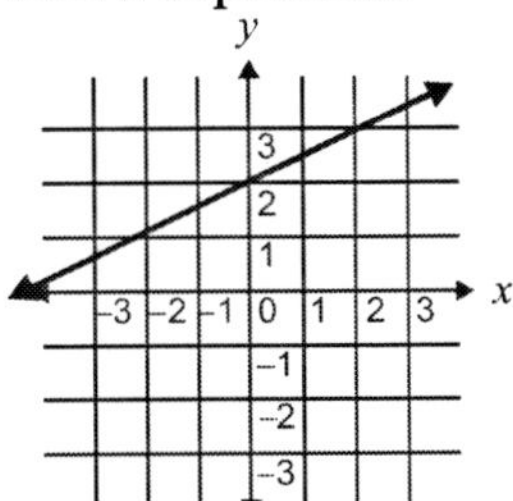

(A) $y = \frac{1}{2}x$

(B) $y = \frac{1}{2}x - 2$

(C) $y = \frac{1}{2}x + 2$

(D) $y = -2x$

4.01a

31. Solve the following quadratic equation by factoring: $3x^2 = 4x + 7$

(A) $-\frac{1}{3}, \frac{7}{3}$

(B) $-1, \frac{7}{3}$

(C) $-1, -2$

(D) $-\frac{7}{3}, \frac{3}{7}$

4.02

32. Multiply and simplify: $(3x + 2)(x - 4)$

(A) $3x^2 - 10x - 8$
(B) $3x^2 + 5x - 8$
(C) $3x^2 + 5x - 6$
(D) $8x^2 - 2$

1.01c

33. Which of the following is a graph of the inequality $y \leq x - 3$?

(A)

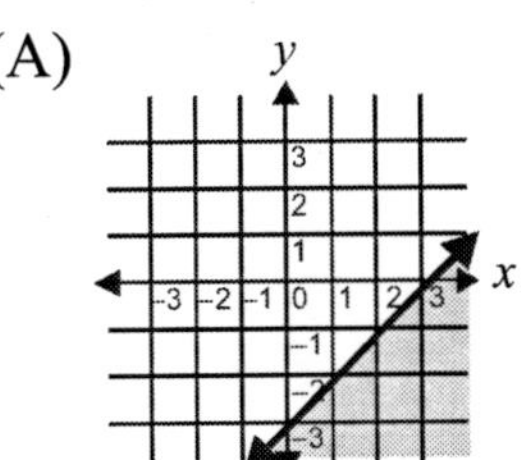

(B)

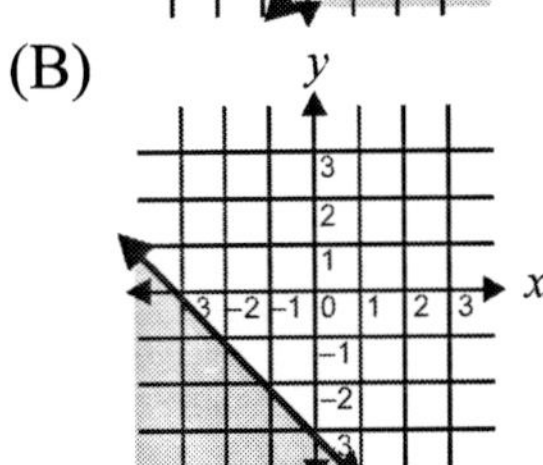

(C)

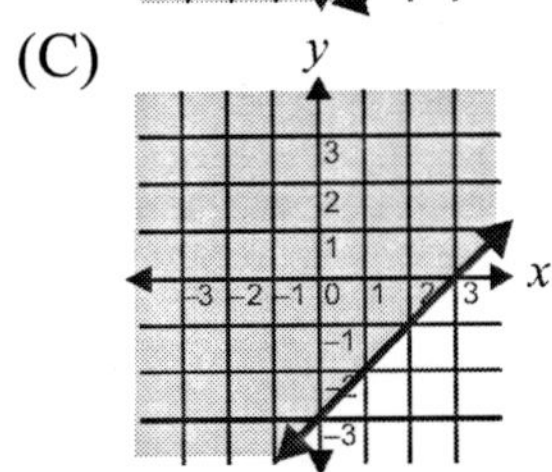

(D)

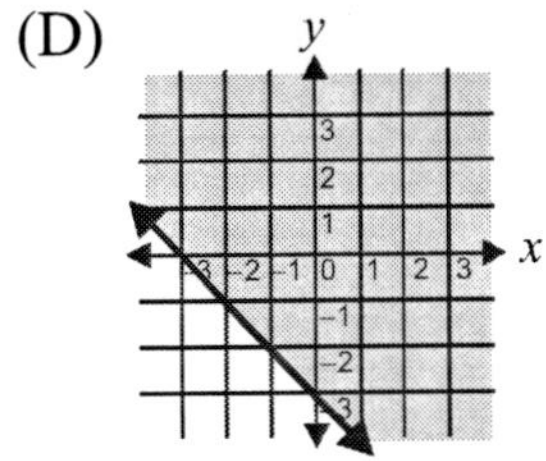

4.01a

34. Which of these is the equation of the line that generalizes the pattern of the data in the table?

x	f(x)
–4	–7
0	1
1	3
6	13

(A) $f(x) = -4 - 7$
(B) $f(x) = x + 2$
(C) $f(x) = 2x + 1$
(D) $f(x) = -7 - 4$

4.01a

35. Which of these expressions represents the average rate in miles per hour between 7 AM and 11 AM?

Time	Odometer Reading
7 AM	20825
11 AM	20965

(A) $\frac{20825 - 20965}{11 - 7}$

(B) $\frac{20825 - 7}{20965 - 11}$

(C) $\frac{20965 - 11}{20965 - 7}$

(D) $\frac{20965 - 20825}{11 - 7}$

3.03a

36. Solve the following inequality:
$-3(4x + 5) > 2(5x + 6) + 13$

(A) $x < -\frac{20}{11}$

(B) $x > 20$

(C) $x > \frac{20}{11}$

(D) $x < 20$

4.01a

37. Which of the following lines is parallel to $y = -4x + 6$?

(A) $y = -2x + 6$
(B) $y = -4x + 2$
(C) $y = 4x + 6$
(D) $y = 2x + 6$

2.02

38. Alexandria wants to locate the midpoint of a line segment with endpoints $(-3, -2)$ and $(6, -4)$. What are the coordinates of the midpoint?

(A) $(1.5, -3)$
(B) $(4.5, -3)$
(C) $(4.5, -6)$
(D) $(3, -6)$

2.01

39. $-2\begin{bmatrix} -5 & -2 \\ 9 & -7 \\ 3 & 6 \end{bmatrix} + \begin{bmatrix} -1 & -2 \\ -8 & 7 \\ 9 & -4 \end{bmatrix} =$

(A) $\begin{bmatrix} -6 & -4 \\ 1 & 0 \\ 12 & 2 \end{bmatrix}$

(B) $\begin{bmatrix} -12 & -8 \\ 1 & 0 \\ 12 & 2 \end{bmatrix}$

(C) $\begin{bmatrix} 9 & 2 \\ -26 & 21 \\ 3 & -16 \end{bmatrix}$

(D) $\begin{bmatrix} 9 & -26 & 3 \\ 2 & 21 & -16 \end{bmatrix}$

3.02

40. Transpiration is the process by which a plant loses water vapor through its leaves.

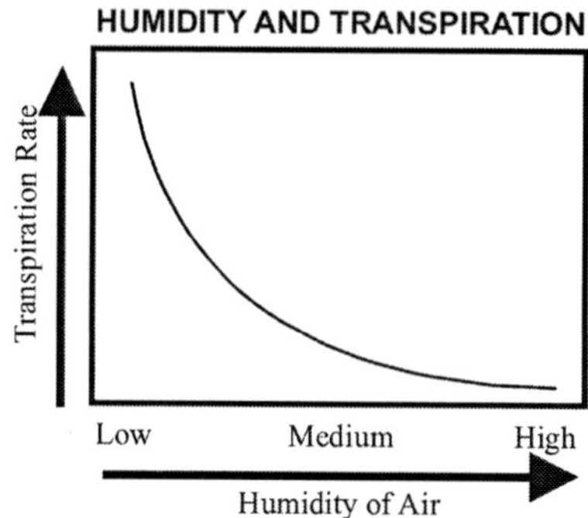

According to the graph, which of the following would be an accurate conclusion?

(A) On days of low humidity, the transpiration rate is highest.
(B) On days of medium humidity, the transpiration rate is above average.
(C) On very humid days, the transpiration rate is highest.
(D) On days of low humidity, the transpiration rate is lowest.

4.04

41. The biology club plans to rent a river boat for an outing. The cost will be \$250.00, plus \$25.00 for each hour (h) the boat is used. Which statement represents the total cost (c) of renting the boat?

(A) $c = 275h$
(B) $c = 250 + 25h$
(C) $c = 275h + 25$
(D) $c = 275 + 25h$

4.01b

42. Brad and Kyle went to the O'Donald's to get a snack after football practice. The table below shows what they bought and the amount they paid.

	Fries	**Drinks**	**Total Cost**
Brad	2	2	\$4.02
Kyle	3	1	\$4.25

What is the cost of 1 order of French fries?

(A) \$0.89
(B) \$1.12
(C) \$1.24
(D) \$1.49

4.03

43. Aunt Bess uses 3 cups of oatmeal to bake 6 dozen oatmeal cookies. How many cups of oatmeal would she need to bake 15 dozen cookies?

(A) 1.2
(B) 7.5
(C) 18
(D) 30

1.03

44. If $c = 5$ and $d = 3$, evaluate $c^2 - 3d$.

(A) 1
(B) 14
(C) 16
(D) 19

4.01a

45. Which of the following algebraic expressions corresponds to:
"the product of 5 and x divided by 3 fewer than y"

(A) $\dfrac{5x}{y-3}$

(B) $\dfrac{5}{x(y-3)}$

(C) $\dfrac{5x}{3-y}$

(D) $\dfrac{5x}{y+3}$

4.01a

46. What would replace n in this number sentence to make the sentence true?
$2n = 16$

(A) 4
(B) 6
(C) 8
(D) 12

4.01b

47. Find v: $v + 3 \geq 24$

(A) $v \leq 21$
(B) $v \leq 27$
(C) $v \geq 21$
(D) $v \geq 27$

4.01a

48. The function rule is $3x(x + 5)$

x	$f(x)$
-2	

(A) -18
(B) 18
(C) -42
(D) 42

4.01a

49. Find the solution to $4m^2 = 9m + 9$.

(A) $\left\{-\frac{3}{2}, \frac{3}{2}\right\}$

(B) $\left\{3, -\frac{3}{4}\right\}$

(C) $\left\{-3, \frac{3}{4}\right\}$

(D) $\left\{-1, \frac{1}{4}\right\}$

4.02

50. Solve the following quadratic equation by factoring: $64x^2 = 25$

(A) $\left\{-\frac{5}{8}, \frac{5}{8}\right\}$

(B) $\left\{\frac{5}{8}, -\frac{8}{5}\right\}$

(C) $\left\{\frac{8}{5}, -\frac{5}{8}\right\}$

(D) $\left\{-\frac{8}{5}, -\frac{5}{8}\right\}$

4.02

51. David earns \$9.50 per hour and works 40 hours each week. He also earns 8% of sales over \$2000 per week. If t represents David's total sales for the week, which equation will help him calculate his earnings from both his hourly wage and his commission for one week?

(A) $y = (9.50 + t)(0.08 + 2000)$
(B) $y = (0.08)(9.50) + (2000 - t)$
(C) $y = (9.50)(40) + (0.08)(t - 2000)$
(D) $y = (9.50)(40) + 8(t - 2000)$

4.01b

52. Two lines are shown on the grid. One line passes through the origin and the other passes through $(-1, -1)$ with a y-intercept of 2. Which pair of equations below the grid identifies these lines?

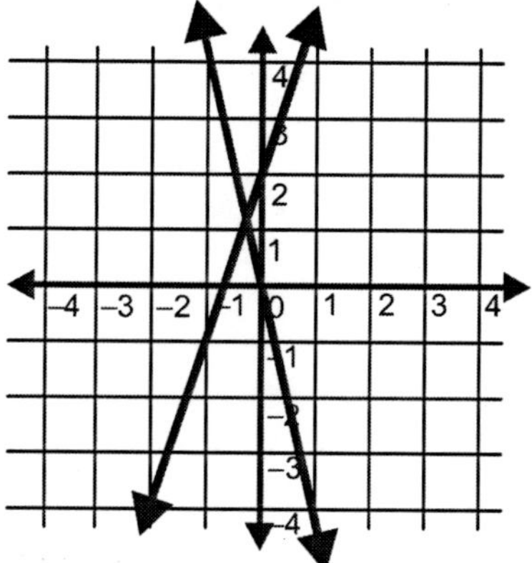

(A) $y = \frac{1}{4}x$ and $y = \frac{1}{3}x + 2$

(B) $x - 2y = 6$ and $4x + y = 4$

(C) $y = 4x$ and $y = \frac{1}{3}x$

(D) $y = 3x + 2$ and $y = -4x$

4.03

53. Using the distance formula, what is the length of $\overline{AB}$?

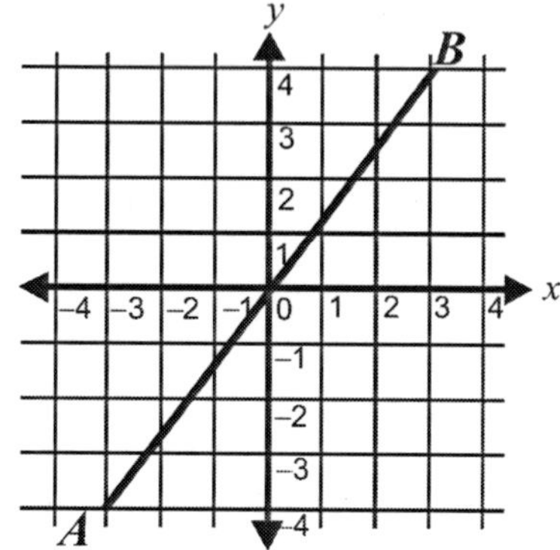

(A) $\sqrt{14}$
(B) 8
(C) 10
(D) 20

2.01

54. Pizza Corner pays its pizza delivery drivers \$48 for a six-hour shift plus \$2 for every pizza delivered. A pizza delivery driver's earnings for one shift at Pizza Corner can be expressed by the following function: $f(x) = 2x + 48$.
Leon's Pizza Palace pays its pizza delivery drivers \$4 for every pizza delivered. A pizza delivery driver's earnings for one shift at Leon's Pizza Palace can be expressed by the following function: $f(x) = 4x$.
How does the graph of earnings at Pizza Corner compare to the graph of earnings at Leon's Pizza Palace?

(A) It has a flatter slope and a smaller y-intercept.
(B) It has a steeper slope and a smaller y-intercept.
(C) It has a flatter slope and a greater y-intercept.
(D) It has a steeper slope and a greater y-intercept.

4.01b

55. A scientist wants to determine the half life of iodine-131 experimentally. He started with 1 gram of iodine-131 and recorded the day when half of the iodine decomposed. He recorded the following data which shows that the mass decayed by half every eight days.

Half-life (n)	Mass
0	1 gram
1	0.5 gram
2	0.25 gram
3	0.125 gram

Which of the following functions describes the decay of mass for iodine-131?

(A) $f(n) = 2^n$
(B) $f(n) = \frac{1}{2}n$
(C) $f(n) = \dfrac{1}{n}$
(D) $f(n) = \dfrac{1}{2^n}$

4.04

56. Which ordered pair is a solution for the following system of equations?

$$-3x + 7y = 25$$
$$3x + 3y = -15$$

(A) $(-13, -2)$
(B) $(-6, 1)$
(C) $(-3, -2)$
(D) $(-20, -5)$

4.03

57. A coin bank contains dimes and quarters. The number of dimes is three less than four times the number of quarters. The total amount in the bank is \$8.15. How many dimes are in the bank?

(A) 13
(B) 41
(C) 49
(D) 65

4.03

58. Solve for y:
$9y^2 - 64 = 0$

(A) $\left\{\frac{8}{9}, -\frac{8}{9}\right\}$
(B) $\left\{\frac{9}{8}, \frac{8}{9}\right\}$
(C) $\left\{\frac{3}{8}, -\frac{3}{8}\right\}$
(D) $\left\{\frac{8}{3}, -\frac{8}{3}\right\}$

4.02

59. What is the distance between points $(-2, -4)$ and $(4, -7)$?

(A) $\sqrt{3}$
(B) 45
(C) 3
(D) $3\sqrt{5}$

2.01

60. Identify the graph of the following function: $y = x^2 - 2$

(A)

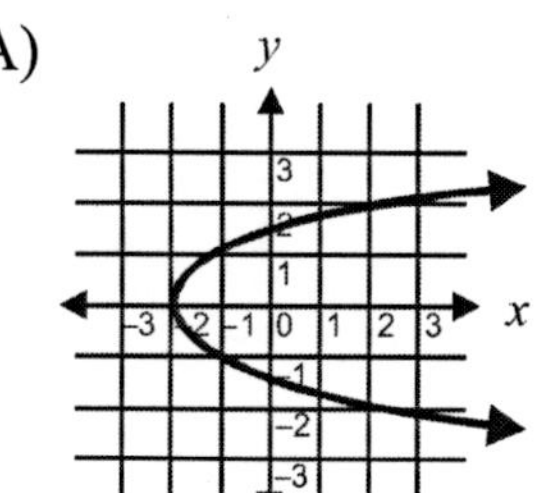

(B)

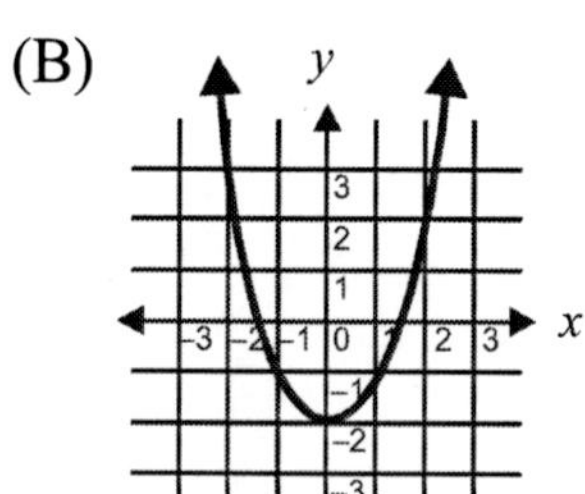

(C)

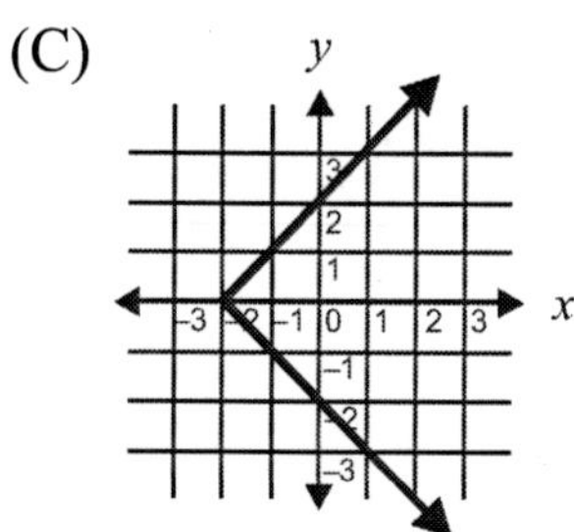

(D)

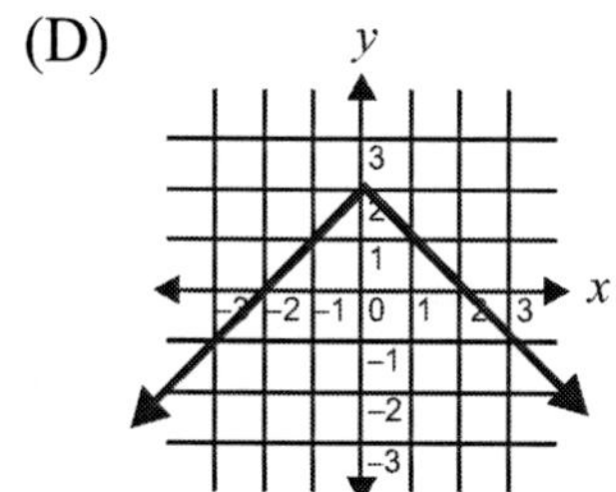

4.02

61. If $2a + b = 13$, then $a =$

(A) $13 - b$

(B) $\frac{13 - b}{2}$

(C) $\frac{b - 13}{2}$

(D) $2(13 - b)$

4.01a

62. What is the slope of the line in the graph below?

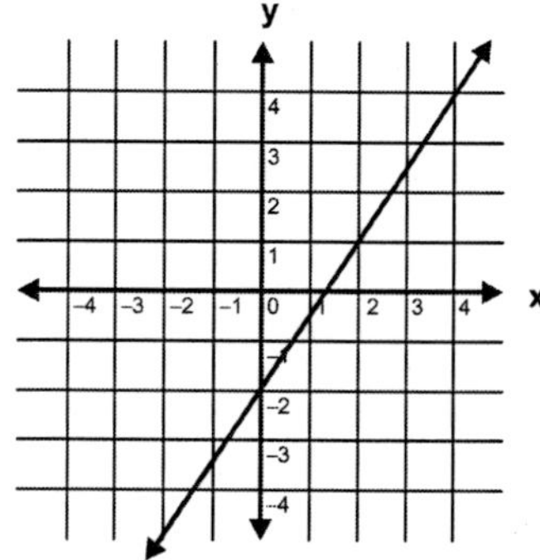

(A) 3

(B) -3

(C) $\frac{2}{3}$

(D) $\frac{3}{2}$

4.01b

63.

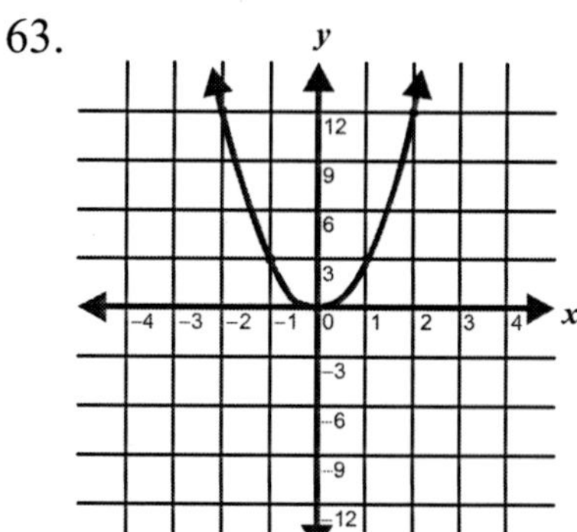

What is the equation graphed above?

(A) $y = 3x^2$

(B) $y = |2x|$

(C) $y = 3x$

(D) $y = 6x$

4.02

64. Find the slope of a line containing the points $(2, 5)$ and $(-2, 5)$. Use the following formula:

slope $= \frac{y_2 - y_1}{x_2 - x_1}$

(A) undefined

(B) 4

(C) -4

(D) 0

4.01b

65. The number of feet any object will fall in a number of seconds is found by the formula:

$s = \frac{1}{2}gt^2$
$s =$ the number of feet an object will fall
$t =$ the number of seconds
$g = 32$ feet per second^2

If an object is dropped from an airplane, how far will it fall in 10 seconds? (Use the formula above.)

(A) 160 feet
(B) 1600 feet
(C) 320 feet
(D) 3200 feet 4.02

66. Where is the y-intercept for the line $3x + 2y + 8 = 0$?

(A) $(2, 0)$
(B) $(0, 8)$
(C) $(0, -4)$
(D) $(-4, 0)$ 4.01b

67. Factor $9x^2 + 15x - 14$.

(A) $(9x - 1)(x + 7)$
(B) $(3x + 2)(3x - 7)$
(C) $(3x - 2)(3x - 7)$
(D) $(3x - 2)(3x + 7)$ 1.01c

68. Solve $a^2 + 5a - 14 = 0$ for a.

(A) The solutions are 2 and -7.
(B) The solutions are -2 and 7.
(C) The solutions are -2 and -7.
(D) The solutions are 2 and 7. 4.02

69. Find y: $-4y < 56$

(A) $y > -14$
(B) $y > 14$
(C) $y < -14$
(D) $y < 14$ 4.01a

70. For the following pair of equations, find the point of intersection (common solution) using the substitution method.

$3x + 3y = 9$
$9y - 3x = 6$

(A) $(1, 2)$
(B) $\left(\frac{7}{4}, \frac{5}{4}\right)$
(C) $(1, 1)$
(D) $\left(\frac{1}{3}, \frac{1}{6}\right)$ 4.03

71. For the following pair of equations, find the point of intersection (common solution) using the substitution method.

$-3x - y = -2$
$5x + 2y = 20$

(A) $(2, -4)$
(B) $(2, 5)$
(C) $(-16, 50)$
(D) $\left(\frac{1}{5}, \frac{1}{2}\right)$ 4.03

72. Solve by factoring:

$11x^2 - 31x - 6 = 0$

(A) 3 and -2
(B) 3 and $-\frac{2}{11}$
(C) 3 and $-\frac{11}{2}$
(D) -3 and $\frac{11}{2}$ 4.02

73. What is the value of A when $t = 3$?
$A = (3)^t$

(A) 1
(B) 9
(C) 27
(D) Cannot be determined. 4.04

74. The local hardware store has four different style grills for the summer sale. The price of the four grills are listed from least to greatest, and they are represented in the 2 x 2 matrix below.

$$\begin{bmatrix} \$120 & \$155 \\ \$160 & \$230 \end{bmatrix}$$

If the store is having a sale where everything is 20% off, what is the new price of the grill that is the second most expensive?

(A) \$32
(B) \$124
(C) \$184
(D) \$128

3.01

75. What is the solution to the systems of equations shown below?

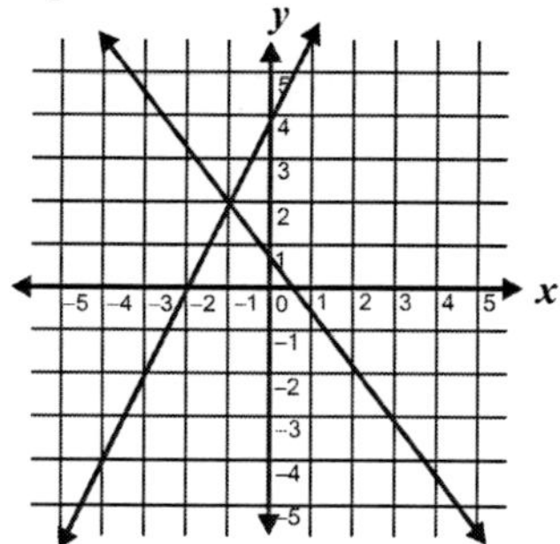

(A) $(2, -1)$
(B) $(-1, 2)$
(C) $x \geq -1$
(D) $y \geq 2$

4.03

76. $-1\begin{bmatrix} 1 & -6 \\ 2 & -3 \end{bmatrix} + 3\begin{bmatrix} 3 & -1 \\ 0 & -3 \end{bmatrix} =$

(A) $\begin{bmatrix} 8 & 3 \\ -2 & -6 \end{bmatrix}$

(B) $\begin{bmatrix} 11 & 6 \\ 1 & 9 \end{bmatrix}$

(C) $\begin{bmatrix} 4 & -7 \\ 2 & -6 \end{bmatrix}$

(D) $\begin{bmatrix} 10 & -9 \\ 2 & -12 \end{bmatrix}$

3.02

77. Which of the following sets of data below can be represented by a linear function?

(A)

x	y
-2	-3
0	3
2	9
4	15

(B)

x	y
-5	-10
-2	-7
2	-1
3	15

(C)

x	y
1	3
2	-2
3	-4
4	-10

(D)

x	y
-2	6
-1	3
1	3
2	6

4.01a

78. The discount store next to the mall is having a huge sale. All pants normally priced \$25.95 are now 40% off, and all shoes normally priced \$14.95 are now 60% off. Which of the following best represents this information in a matrix?

(A) $\begin{bmatrix} \$25.95 & 40\% \\ \$14.95 & 60\% \end{bmatrix}$

(B) $\begin{bmatrix} \$25.95 & 60\% \\ \$14.95 & 40\% \end{bmatrix}$

(C) $\begin{bmatrix} \$25.95 & \$14.95 \\ 40\% & 60\% \end{bmatrix}$

(D) Both A and C represent the data correctly.

3.03b

79. Using the following function, what is the value of $f(x)$ when $x = 3$? $f(x) = 2x - 5$

(A) 1
(B) 4
(C) -4
(D) -1

4.01a

80. $\begin{bmatrix} 1 & 3 \\ 1 & -4 \end{bmatrix} + 3 \begin{bmatrix} 3 & -1 & 0 \\ 6 & 5 & 2 \end{bmatrix} =$

(A) $\begin{bmatrix} 10 & 0 & 0 \\ 19 & 11 & 6 \end{bmatrix}$

(B) $\begin{bmatrix} 10 & -3 & 0 \\ 19 & 15 & 2 \end{bmatrix}$

(C) $\begin{bmatrix} 10 & 0 \\ 19 & 11 \end{bmatrix}$

(D) Not possible

3.02

Evaluation Chart for the Diagnostic Mathematics Test

Directions: On the following chart, circle the question numbers that you answered incorrectly. Then turn to the appropriate topics (listed by chapters), read the explanations, and complete the exercises. Review the other chapters as needed. Finally, complete the *Passing the North Carolina Algebra I End-of-Course* Practice Tests to further review.

		Questions	Pages
Chapter 1:	Numbers and Exponents	1, 8	17–27
Chapter 2:	Introduction to Algebra	3, 16, 17, 24, 28, 41, 44, 45, 51	28–36
Chapter 3:	Solving Multi-Step Equations and Inequalities	7, 10, 13, 14, 36, 46, 47, 61, 69	37–52
Chapter 4:	Ratios and Proportions	4, 43	53–59
Chapter 5:	Algebra Word Problems		60–70
Chapter 6:	Matrices	39, 74, 76, 78, 80	71–77
Chapter 7:	Polynomials	2, 15	78–88
Chapter 8:	Factoring	32, 67	89–99
Chapter 9:	Solving Quadratic Equations	19, 23, 25, 26, 31, 49, 50, 58, 65, 68, 72	100–106
Chapter 10:	Graphing and Writing Equations and Inequalities	5, 22, 33, 38, 53, 59, 62, 64, 66	107–127
Chapter 11:	Applications of Graphs	6, 9, 11, 18, 21, 30, 34, 35, 54, 60, 63	128–148
Chapter 12:	Systems of Equations and Systems of Inequalities	12, 27, 29, 37, 42, 52, 56, 57, 70, 71, 75	149–156
Chapter 13:	Relations and Functions	20, 40, 48, 55, 73, 77, 79	157–172

Correlation of Chapters to Standards

The following chart lists every chapter in this book. Each chapter is benchmarked to the North Carolina 2003 Standard Course of Study as listed on the North Carolina Department of Education website.

Note to educators: To cover a specific standard, find the standard on the chart and turn to the chapter it is correlated to. Some standards are covered in more than one chapter.

	Competency Goal	Objective	Pages
Chapter 1: Numbers and Exponents	Number and Operations	1.01a	17–27
Chapter 2: Introduction to Algebra	Number and Operations	1.01 1.02	28–36
	Algebra	4.01a, b	
Chapter 3: Solving Multi-Step Equations and Inequalities	Number and Operations	1.01 1.02	37–52
	Algebra	4.01a, b	
Chapter 4: Ratios and Proportions	Number and Operations	1.01 1.02 1.03	53–59
	Algebra	4.01	
Chapter 5: Algebra Word Problems	Number and Operations	1.01 1.02	60–70
	Algebra	4.01	
Chapter 6: Matrices	Number and Operations	1.01 1.02	71–77
	Data Analysis and Probability	3.01 3.02 3.03b	
	Algebra	4.01	
Chapter 7: Polynomials	Number and Operations	1.01b	78–88
Chapter 8: Factoring	Number and Operations	1.01b, c	89–99
Chapter 9: Solving Quadratic Equations	Algebra	4.02	100–106

		Competency Goal	Objective	Pages
Chapter 10:	Graphing and Writing Equations and Inequalities	Number and Operations	1.01	107–127
			1.02	
		Geometry and Measurement	2.01	
		Data Analysis and Probability	3.03a, b	
		Algebra	4.01a, b	
Chapter 11:	Applications of Graphs	Number and Operations	1.01	128–148
			1.02	
		Geometry and Measurement	2.02	
		Data Analysis and Probability	3.03a, b	
		Algebra	4.01a	
			4.02	
Chapter 12:	Systems of Equations and Systems of Inequalities	Number and Operations	1.01	149–156
			1.02	
		Geometry and Measurement	2.02	
		Algebra	4.01a	
			4.03	
Chapter 13:	Relations and Functions	Number and Operations	1.01	157–172
			1.02	
		Algebra	4.01a	
			4.04	

Chapter 1
Numbers and Exponents

This chapter covers the following North Carolina mathematics standards for Algebra I:

Competency Goal	Objective
Number and Operations	1.01a

1.1 Real Numbers

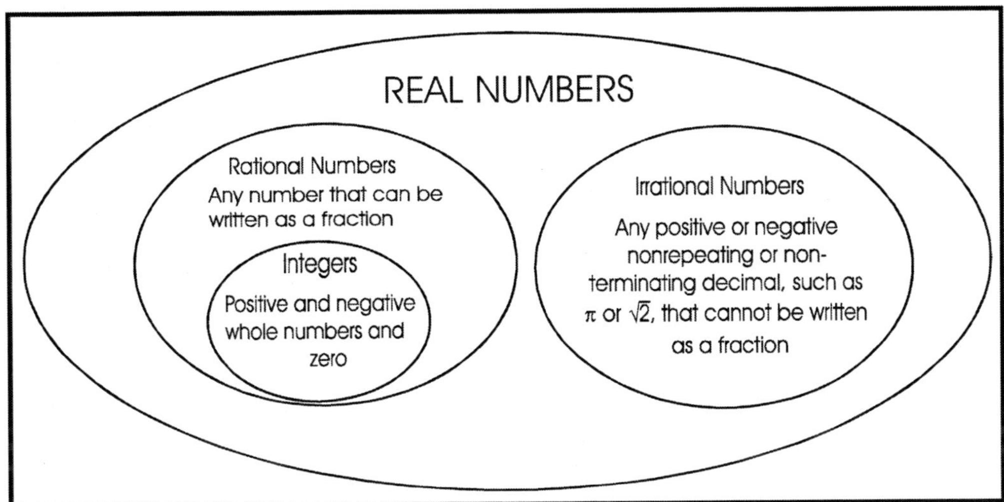

Real numbers include all positive and negative numbers and zero. Included in the set of real numbers are positive and negative fractions, decimals, and rational and irrational numbers.

Use the diagram above and your calculator to answer the following questions.

1. Using your calculator, find the square root of 7. Does it repeat? Does it end? Is it a rational or an irrational number?
2. Find $\sqrt{25}$. Is it rational or irrational? Is it an integer?
3. Is an integer an irrational number?
4. Is an integer a real number?
5. Is $\frac{1}{8}$ a real number? Is it rational or irrational?

Identity the following numbers as rational (R) or (I).

6. 5π
7. $\sqrt{8}$
8. $\frac{1}{3}$
9. -7.2
10. $-\frac{3}{4}$
11. $\frac{\sqrt{2}}{2}$
12. $9 + \pi$
13. 1.0004
14. $-\frac{4}{5}$
15. $1.1\overline{8}$
16. $\sqrt{81}$
17. $\frac{\pi}{4}$
18. $-\sqrt{36}$
19. $17\frac{1}{2}$
20. $-\frac{5}{3}$

1.2 Integers

In elementary school, you learned to use whole numbers.

$$\textbf{Whole numbers} = \{0, 1, 2, 3, 4, 5, ...\}$$

For most things in life, whole numbers are all we need to use. However, when a checking account falls below zero or the temperature falls below zero, we need a way to express that. Mathematicians have decided that a negative sign, which looks exactly like a subtraction sign, would be used in front of a number to show that the number is below zero. All the negative whole numbers and positive whole numbers plus zero make up the set of integers.

$$\textbf{Integers} = \{..., -4, -3, -2, -1, 0, 1, 2, 3, 4, ...\}$$

1.3 Absolute Value

The absolute value of a number is the distance the number is from zero on the number line.

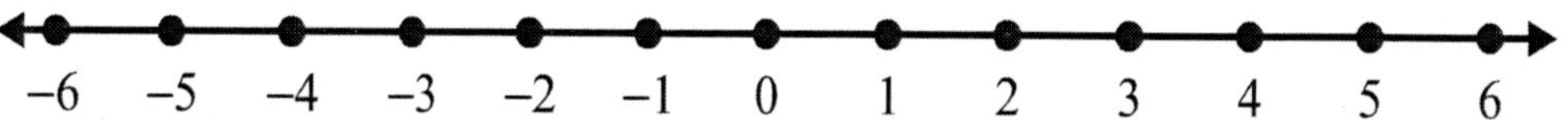

The absolute value of 6 is written $\|6\|$.	$\|6\| = 6$
The absolute value of -6 is written $\|-6\|$.	$\|-6\| = 6$

Both 6 and -6 are the same distance, 6 spaces, from zero so their absolute value is the same: 6.

Examples:

$\|-4\| = 4$	$-\|-4\| = -4$	$\|-9\| + 5 = 9 + 5 = 14$
$\|9\| - \|8\| = 9 - 8 = 1$	$\|6\| - \|-6\| = 6 - 6 = 0$	$\|-5\| + \|-2\| = 5 + 2 = 7$

Simplify the following absolute value problems.

1. $\|9\| =$ ______	6. $\|-2\| =$ ______	11. $\|-2\| + \|6\| =$ ______
2. $-\|5\| =$ ______	7. $-\|-3\| =$ ______	12. $\|10\| + \|8\| =$ ______
3. $\|-25\| =$ ______	8. $\|-4\| - \|3\| =$ ______	13. $\|-2\| + \|4\| =$ ______
4. $-\|-12\| =$ ______	9. $\|-8\| - \|-4\| =$ ______	14. $\|-3\| + \|-4\| =$ ______
5. $-\|64\| =$ ______	10. $\|5\| + \|-4\| =$ ______	15. $\|7\| - \|-5\| =$ ______

1.4 Understanding Exponents

Sometimes it is necessary to multiply a number by itself one or more times. For example, a math problem may need to multiply 3×3 or $5 \times 5 \times 5 \times 5$. In these situations, mathematicians have come up with a shorter way of writing out this kind of multiplication. Instead of writing 3×3, you can write 3^2, or instead of writing $5 \times 5 \times 5 \times 5$, 5^4 means the same thing. The first number is the **base**. The small, raised number is called the **exponent** or **power**. The exponent tells how many times the base should be multiplied by itself.

Example 1: 6^3 ← **exponent (or power)**; 6 ← **base**. This means multiply by 6 three times: $6 \times 6 \times 6$

Example 2: **Negative numbers can be raised to exponents also.**
An **even** exponent will give a **positive** answer: $(-2)^2 = (-2) \times (-2) = 4$
An **odd** exponent will give a **negative** answer: $(-2)^3 = (-2) \times (-2) \times (-2) = (-8)$

You also need to know two special properties of exponents:

1. **Any base number raised to the exponent of 1 equals the base number.**
2. **Any base number raised to the exponent of 0 equals 1.**

Example 3: $4^1 = 4$ $10^1 = 10$ $25^1 = 25$ $4^0 = 1$ $10^0 = 1$ $25^0 = 1$

Rewrite the following problems using exponents.

Example 4: $2 \times 2 \times 2 = 2^3$

1. $7 \times 7 \times 7 \times 7$
2. 10×10
3. $12 \times 12 \times 12$
4. $4 \times 4 \times 4 \times 4$
5. $9 \times 9 \times 9$
6. 25×25
7. $15 \times 15 \times 15$
8. $5 \times 5 \times 5 \times 5 \times 5$
9. $2 \times 2 \times 2 \times 2$
10. 14×14
11. $3 \times 3 \times 3 \times 3 \times 3$
12. $11 \times 11 \times 11$

Use your calculator to figure what product each number with an exponent represents.

Example 5: $2^3 = 2 \times 2 \times 2 = 8$

13. $(-8)^3$
14. 12^2
15. 20^1
16. 5^4
17. 15^0
18. 16^2
19. $(-10)^2$
20. 3^5
21. 10^4
22. 7^0
23. 4^3
24. 54^1

Express each of the following numbers as a base with an exponent.

Example 6: $4 = 2 \times 2 = 2^2$

25. 9
26. 16
27. 27
28. 36
29. 8
30. 32
31. 1000
32. 125
33. 81
34. 64
35. 49
36. 121

1.5 Multiplication with Exponents

Rule 1: To multiply two expressions with the same base, add the exponents together and keep the base the same.

Example 7: $2^3 \times 2^5 = 2^{3+5} = 2^8$

Rule 2: If a power is raised to another power, multiply the exponents together and keep the base the same.

Example 8: $(2^3)^2 = 2^{3\times2} = 2^6$

Rule 3: If a product in parenthesis is raised to a power, then each factor is raised to the power when parenthesis are eliminated.

Example 9: $(2 \times 4)^2 = 2^2 \times 4^2 = 4 \times 16 = 64$
Example 10: $(3a)^3 = 3^3 \times a^3 = 27a^3$
Example 11: $(7b^5)^2 = 7^2b^{10} = 49b^{10}$

Simplify each of the expressions below.

1. $(5^3)^2$
2. $6^3 \times 6^5$
3. $4^3 \times 4^3$
4. $(7^5)^2$
5. $(6^2)^5$
6. $2^5 \times 2^3$
7. $(4 \times 5)^2$
8. $(3^4)^0$
9. $(3^3)^2$
10. $2^5 \times 2^5$
11. $(3 \times 3)^2$
12. $(2a)^4$
13. $(3^2)^4$
14. $4^5 \times 4^3$
15. $(3 \times 2)^4$
16. $(5^2)^2$
17. $(6 \times 4)^2$
18. $(9a^5)^3$
19. $4^3 \times 4^4$
20. $(6b^5)^2$
21. $(5^2)^3$
22. $3^7 \times 3^3$
23. $(3a)^2$
24. $(3^4)^2$
25. $(4^4)^2$
26. $(2b^3)^4$
27. $(5a^2)^5$
28. $(8a^3)^2$
29. $(9^2)^2$
30. $10^5 \times 10^4$
31. $(3 \times 5)^2$
32. $(7^3)^2$

1.6 Division with Exponents

Rule 1: Expressions can also have negative exponents. Negative exponents do not indicate negative numbers. They indicate reciprocals, which is 1 over the original number.

Example 12: $2^{-3} = \frac{1}{2^3} = \frac{1}{8}$

Example 13: $3a^{-5} = 3 \times \frac{1}{a^5} = \frac{3}{a^5}$

Rule 2: When dividing expressions with exponents that have the same base, subtract the exponents. Expressions in simplified form only have positive exponents.

Example 14: $\frac{3^5}{3^3} = 3^{5-3} = 3^2 = 9$

Example 15: $\frac{3^5}{3^8} = 3^{5-8} = 3^{-3} = \frac{1}{3^3} = \frac{1}{27}$

Rule 3: If a fraction is raised to a power, then both the numerator and the denominator are raised to the same power.

Example 16: $\left(\frac{3}{4}\right)^3 = \frac{3^3}{4^3} = \frac{27}{64}$

Example 17: $(2x)^{-2} = \frac{1}{(2x)^2} = \frac{1}{4x^2}$

Reduce the following expressions to their simplest form. All exponents should be positive.

1. $5x^{-4}$
2. $\frac{2^2}{2^4}$
3. $\left(\frac{2}{3}\right)^2$
4. $6a^{-2}$
5. $\frac{3^6}{3^3}$
6. $(5a)^{-2}$
7. $\frac{3^4}{3^3}$
8. $\left(\frac{7}{8}\right)^3$
9. $(6a)^{-2}$
10. $\frac{(x^2)^3}{x^4}$
11. $\frac{(3y)^3}{3^2y}$
12. $\frac{(3a^2)^3}{a^4}$
13. $(2x^2)^{-5}$
14. $2x^{-2}$
15. $(a^3)^{-2}$
16. $(2^{-2})^3$
17. $\left(\frac{1}{2}\right)^2$
18. $\frac{1}{3^{-2}}$
19. $(4y)^{-5}$
20. $4y^{-5}$

1.7 Square Root

Just as working with exponents is related to multiplication, so finding square roots is related to division. In fact, the sign for finding the square root of a number looks similar to a division sign. The best way to learn about square roots is to look at examples.

Example 18: This is a square root problem: $\sqrt{64}$
It is asking, "What is the square root of 64?"
It means, "What number multiplied by itself equals 64?"
The answer is 8. $8 \times 8 = 64$.

Example 19: Find the square root of 36 and 144.

$\sqrt{36}$
$6 \times 6 = 36$ so $\sqrt{36} = 6$

$\sqrt{144}$
$12 \times 12 = 144$ so $\sqrt{144} = 12$

Find the square roots of the following numbers.

1. $\sqrt{49}$
2. $\sqrt{81}$
3. $\sqrt{25}$
4. $\sqrt{16}$
5. $\sqrt{121}$
6. $\sqrt{625}$
7. $\sqrt{100}$
8. $\sqrt{289}$
9. $\sqrt{196}$
10. $\sqrt{36}$
11. $\sqrt{4}$
12. $\sqrt{900}$
13. $\sqrt{64}$
14. $\sqrt{9}$
15. $\sqrt{144}$

1.8 Adding and Subtracting Roots

You can add and subtract terms with square roots only if the number under the square root sign is the same.

Example 20: $2\sqrt{2} + 3\sqrt{2} = 5\sqrt{2}$

Example 21: $12\sqrt{7} - 3\sqrt{7} = 9\sqrt{7}$

Or, look at the following examples where you can simplify the square roots and then add or subtract.

Example 22: $2\sqrt{25} + \sqrt{36}$

Step 1: Simplify. You know that $\sqrt{25} = 5$, and $\sqrt{36} = 6$ so the problem simplifies to $2(5) + 6$

Step 2: Solve: $2(5) + 6 = 10 + 6 = 16$

Example 23: $2\sqrt{72} - 3\sqrt{2}$

Step 1: Simplify what you know. $\sqrt{72} = \sqrt{36 \cdot 2} = 6\sqrt{2}$

Step 2: Substitute $6\sqrt{2}$ for $\sqrt{72}$ simplify.
$2(6)\sqrt{2} - 3\sqrt{2} = 12\sqrt{2} - 3\sqrt{2} = 9\sqrt{2}$

Simplify the following addition and subtraction problems.

1. $3\sqrt{5} + 9\sqrt{5}$
2. $3\sqrt{25} + 4\sqrt{16}$
3. $4\sqrt{8} + 2\sqrt{2}$
4. $3\sqrt{32} - 2\sqrt{2}$
5. $\sqrt{25} - \sqrt{49}$
6. $2\sqrt{5} + 4\sqrt{20}$
7. $5\sqrt{8} - 3\sqrt{72}$
8. $\sqrt{27} + 3\sqrt{27}$
9. $3\sqrt{20} - 4\sqrt{45}$
10. $4\sqrt{45} - \sqrt{75}$
11. $2\sqrt{28} + 2\sqrt{7}$
12. $\sqrt{64} + \sqrt{81}$
13. $5\sqrt{54} - 2\sqrt{24}$
14. $\sqrt{32} + 2\sqrt{50}$
15. $2\sqrt{7} + 4\sqrt{63}$
16. $8\sqrt{2} + \sqrt{8}$
17. $2\sqrt{8} - 4\sqrt{32}$
18. $\sqrt{36} + \sqrt{100}$
19. $\sqrt{9} + \sqrt{25}$
20. $\sqrt{64} - \sqrt{36}$
21. $\sqrt{75} + \sqrt{108}$
22. $\sqrt{81} + \sqrt{100}$
23. $\sqrt{192} - \sqrt{75}$
24. $3\sqrt{5} + \sqrt{245}$

1.9 Multiplying Roots

You can also multiply square roots. To multiply square roots, you just multiply the numbers under the square root sign and then simplify. Look at the examples below.

Example 24: $\sqrt{2} \times \sqrt{6}$

Step 1: $\sqrt{2} \times \sqrt{6} = \sqrt{2 \times 6} = \sqrt{12}$ Multiply the numbers under the square root sign.

Step 2: $\sqrt{12} = \sqrt{4 \times 3} = 2\sqrt{3}$ Simplify.

Example 25: $3\sqrt{3} \times 5\sqrt{6}$

Step 1: $(3 \times 5)\sqrt{3 \times 6} = 15\sqrt{18}$ Multiply the numbers in front of the square root, and multiply the numbers under the square root sign.

Step 2: $15\sqrt{18} = 15\sqrt{2 \times 9}$
$15 \times 3\sqrt{2} = 45\sqrt{2}$ Simplify.

Example 26: $\sqrt{14} \times \sqrt{42}$ For this more complicated multiplication problem, use the rule of roots that you learned above, $\sqrt{a \cdot b} = \sqrt{a} \cdot \sqrt{b}$.

Step 1: $\sqrt{14} = \sqrt{7} \times \sqrt{2}$ and
$\sqrt{42} = \sqrt{2} \times \sqrt{3} \times \sqrt{7}$ Instead of multiplying 14 by 42, divide these numbers into their roots.

$\sqrt{14} \times \sqrt{42} = \sqrt{7} \times \sqrt{2} \times \sqrt{2} \times \sqrt{3} \times \sqrt{7}$

Step 2: Since you know that $\sqrt{7} \times \sqrt{7} = 7$ and $\sqrt{2} \times \sqrt{2} = 2$, the problem simplifies to $(7 \times 2)\sqrt{3} = 14\sqrt{3}$

Simplify the following multiplication problems.

1. $\sqrt{5} \times \sqrt{7}$
2. $\sqrt{32} \times \sqrt{2}$
3. $\sqrt{10} \times \sqrt{14}$
4. $2\sqrt{3} \times 3\sqrt{6}$
5. $4\sqrt{2} \times 2\sqrt{10}$
6. $\sqrt{5} \times 3\sqrt{15}$
7. $\sqrt{45} \times \sqrt{27}$
8. $5\sqrt{21} \times \sqrt{7}$
9. $\sqrt{42} \times \sqrt{21}$
10. $4\sqrt{3} \times 2\sqrt{12}$
11. $\sqrt{56} \times \sqrt{24}$
12. $\sqrt{11} \times 2\sqrt{33}$
13. $\sqrt{13} \times \sqrt{26}$
14. $2\sqrt{2} \times 5\sqrt{5}$
15. $\sqrt{6} \times \sqrt{12}$

1.10 Dividing Roots

When dividing a number or a square root by another square root, you cannot leave the square root sign in the denominator (the bottom number) of a fraction. You must simplify the problem so that the square root is not in the denominator. Look at the examples below.

Example 27: $\dfrac{\sqrt{2}}{\sqrt{5}}$

Step 1: $\dfrac{\sqrt{2}}{\sqrt{5}} \times \dfrac{\sqrt{5}}{\sqrt{5}} \longleftarrow$ The fraction $\frac{\sqrt{5}}{\sqrt{5}}$ is equal to 1, and multiplying by 1 does not change the value of a number.

Step 2: $\dfrac{\sqrt{2 \times 5}}{5} = \dfrac{\sqrt{10}}{5}$ Multiply and simplify. Since $\sqrt{5} \times \sqrt{5}$ equals 5, you no longer have a square root in the denominator.

Example 28: $\dfrac{6\sqrt{2}}{2\sqrt{10}}$ In this problem, the numbers outside of the square root will also simplify.

Step 1: $\dfrac{6}{2} = 3$ so you have $\dfrac{3\sqrt{2}}{\sqrt{10}}$

Step 2: $\dfrac{3\sqrt{2}}{\sqrt{10}} \times \dfrac{\sqrt{10}}{\sqrt{10}} = \dfrac{3\sqrt{2 \times 10}}{10} = \dfrac{3\sqrt{20}}{10}$

Step 3: $\dfrac{3\sqrt{20}}{10}$ will further simplify because $\sqrt{20} = 2\sqrt{5}$, so you then have $\dfrac{3 \times 2\sqrt{5}}{10}$ which reduces to $\dfrac{3\sqrt{5}}{5}$.

Simplify the following division problems.

1. $\dfrac{9\sqrt{3}}{\sqrt{5}}$
2. $\dfrac{16}{\sqrt{8}}$
3. $\dfrac{24\sqrt{10}}{12\sqrt{3}}$
4. $\dfrac{\sqrt{121}}{\sqrt{6}}$
5. $\dfrac{\sqrt{40}}{\sqrt{90}}$
6. $\dfrac{33\sqrt{15}}{11\sqrt{2}}$
7. $\dfrac{\sqrt{32}}{\sqrt{12}}$
8. $\dfrac{\sqrt{11}}{\sqrt{5}}$
9. $\dfrac{\sqrt{2}}{\sqrt{6}}$
10. $\dfrac{2\sqrt{7}}{\sqrt{14}}$
11. $\dfrac{5\sqrt{2}}{4\sqrt{8}}$
12. $\dfrac{4\sqrt{21}}{7\sqrt{7}}$
13. $\dfrac{9\sqrt{22}}{2\sqrt{2}}$
14. $\dfrac{\sqrt{35}}{2\sqrt{14}}$
15. $\dfrac{\sqrt{40}}{\sqrt{15}}$
16. $\dfrac{\sqrt{3}}{\sqrt{12}}$

1.11 Estimating Square Roots

Example 29: Estimate the value of $\sqrt{3}$.

Step 1: Estimate the value of $\sqrt{3}$ by using the square root of values that you know. $\sqrt{1}$ is 1 and $\sqrt{4}$ is 2, so the value of $\sqrt{3}$ is going to be between 1 and 2.

Step 2: To estimate a little closer, try squaring 1.5. $1.5 \times 1.5 = 2.25$, so $\sqrt{3}$ has to be greater than 1.5. If you do further trial-and-error calculations, you will find that $\sqrt{3}$ is greater than 1.7 ($1.7 \times 1.7 = 2.89$) but less than 1.8 ($1.8 \times 1.8 = 3.24$).

Therefore $\sqrt{3}$ is around 1.75. It is closer to 2 than it is to 1.

Example 30: Is the $\sqrt{52}$ closer to 7 or 8? Look at the perfect square above and below 52.

To answer this question, first look at 7^2 which is equal to 49 and 8^2 which is equal to 64. Then ask yourself whether 52 is closer to 49 or 64. The answer is 49, of course. Therefore, the $\sqrt{52}$ is closer to 7 than 8.

Follow the steps above to answer the following questions. Do not use a calculator.

1. Is $\sqrt{66}$ closer to 8 or 9?
2. Is $\sqrt{27}$ closer to 5 or 6?
3. Is $\sqrt{13}$ closer to 3 or 4?
4. Is $\sqrt{78}$ closer to 8 or 9?
5. Is $\sqrt{12}$ closer to 3 or 4?
6. Is $\sqrt{8}$ closer to 2 or 3?
7. Is $\sqrt{20}$ closer to 4 or 5?
8. Is $\sqrt{53}$ closer to 7 or 8?
9. Is $\sqrt{60}$ closer to 7 or 8?
10. Is $\sqrt{6}$ closer to 2 or 3?

Chapter 1 Review

Simplify the following problems.

1. 15^0
2. $\sqrt{100}$
3. $\sqrt{49}$
4. $(-3)^3$
5. $5^2 \times 5^3$
6. $(4^4)^3$
7. $(3a^2)^{-2}$
8. $6x^{-3}$
9. $\dfrac{4^6}{4^4}$
10. $\left(\dfrac{3}{5}\right)^2$
11. $\dfrac{(3a^2)^3}{a^3}$
12. $\dfrac{6x^{-2}}{x^{-3}}$

Write as exponents.

13. $3 \times 3 \times 3 \times 3$
14. $6 \times 6 \times 6 \times 6 \times 6 \times 6$
15. $11 \times 11 \times 11$
16. $2 \times 2 \times 2 \times 2 \times 2 \times 2 \times 2 \times 2$

Solve the following absolute value problems.

17. $|4|$
18. $|-6|$
19. $|-3| + |7|$
20. $|8| - |-5|$

Simplify the following square root problems.

21. $5\sqrt{27} + 7\sqrt{3}$
22. $\sqrt{40} - \sqrt{10}$
23. $\sqrt{64} + \sqrt{81}$
24. $\dfrac{\sqrt{56}}{\sqrt{35}}$
25. $14\sqrt{5} + 8\sqrt{80}$
26. $\sqrt{63} \times \sqrt{28}$
27. $8\sqrt{50} - 3\sqrt{32}$
28. $\sqrt{8} \times \sqrt{50}$
29. $\dfrac{\sqrt{20}}{\sqrt{45}}$
30. $5\sqrt{40} \times 3\sqrt{20}$
31. $2\sqrt{48} - \sqrt{12}$
32. $\sqrt{72} \times 3\sqrt{27}$
33. $4\sqrt{5} + 8\sqrt{45}$
34. $\dfrac{3\sqrt{22}}{2\sqrt{3}}$
35. $\dfrac{2\sqrt{5}}{\sqrt{30}}$

Estimate the following square root solutions.

36. Is $\sqrt{5}$ closer to 2 or 3?
37. Is $\sqrt{52}$ closer to 7 or 8?
38. Is $\sqrt{130}$ closer to 11 or 12?
39. Is $\sqrt{619}$ closer to 24 or 25?

Chapter 2
Introduction to Algebra

This chapter covers the following North Carolina mathematics standards for Algebra I:

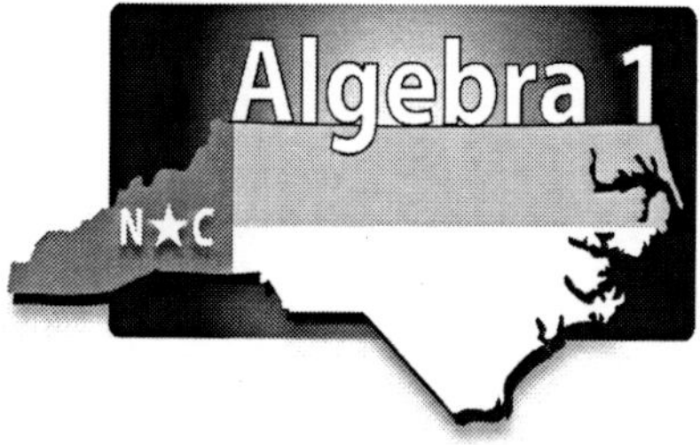

Competency Goal	Objectives
Number and Operations	1.01 1.02
Algebra	4.01a, b

2.1 Algebra Vocabulary

Vocabulary Word	Example	Definition
variable	$4x$ (x is the variable)	a letter that can be replaced by a number
coefficient	$4x$ (4 is the coefficient)	a number multiplied by a variable or variables
term	$5x^2 + x - 2$ ($5x^2$, x, and -2 are terms)	numbers or variables separated by $+$ or $-$ signs
constant	$5x + 2y + 4$ (4 is a constant)	a term that does not have a variable
degree	$4x^2 + 3x - 2$ (the degree is 2)	the largest power of a variable in an expression
leading coefficient	$4x^2 + 3x - 2$ (4 is the leading coefficient)	the number multiplied by the term with the highest power
sentence	$2x = 7$ or $5 \leq x$	two algebraic expressions connected by $=$, $\neq$, $<$, $>$, $\leq$, $\geq$, or $\approx$
equation	$4x = 8$	a sentence with an equal sign
inequality	$7x < 30$ or $x \neq 6$	a sentence with one of the following signs: $\neq$, $<$, $>$, $\leq$, $\geq$, or $\approx$
base	6^3 (6 is the base)	the number used as a factor
exponent	6^3 (3 is the exponent)	the number of times the base is multiplied by itself

2.2 Substituting Numbers for Variables

These problems may look difficult at first glance, but they are very easy. Simply replace the variable with the number the variable is equal to, and solve the problems.

Example 1: In the following problems, substitute 10 for a.

Problem	Calculation	Solution
1. $a + 1$	Simply replace the a with 10. $10 + 1$	11
2. $17 - a$	$17 - 10$	7
3. $9a$	This means multiply. 9×10	90
4. $\frac{30}{a}$	This means divide. $30 \div 10$	3
5. a^3	$10 \times 10 \times 10$	1000
6. $5a + 6$	$(5 \times 10) + 6$	56

Note: Be sure to do all multiplying and dividing before adding and subtracting.

Example 2: In the following problems, let $x = 2$, $y = 4$, and $z = 5$.

Problem	Calculation	Solution
1. $5xy + z$	$5 \times 2 \times 4 + 5$	45
2. $xz^2 + 5$	$2 \times 5^2 + 5 = 2 \times 25 + 5$	55
3. $\frac{yz}{x}$	$(4 \times 5) \div 2 = 20 \div 2$	10

In the following problems, $t = 7$. Solve the problems.

1. $t + 3 =$
2. $18 - t =$
3. $\frac{21}{t} =$
4. $3t - 5 =$
5. $t^2 + 1 =$
6. $2t - 4 =$
7. $9t \div 3 =$
8. $\frac{t^2}{7} =$
9. $5t + 6 =$
10. $\frac{(t^2 - 7)}{6} =$
11. $4t + 5t =$
12. $\frac{6t}{3} =$

In the following problems $a = 4$, $b = -2$, $c = 5$, and $d = 10$. Solve the problems.

13. $4a + 2c =$
14. $3bc - d =$
15. $\frac{ac}{d} =$
16. $d - 2a =$
17. $a^2 - b =$
18. $abd =$
19. $5c - ad =$
20. $cd + bc =$
21. $\frac{6b}{a} =$
22. $9a + b =$
23. $5 + 3bc =$
24. $d^2 + d + 1 =$

2.3 Understanding Algebra Word Problems

The biggest challenge to solving word problems is figuring out whether to add, subtract, multiply, or divide. Below is a list of key words and their meanings. This list does not include every situation you might see, but it includes the most common examples.

Words Indicating Addition	Example	Add
and	6 **and** 8	$6+8$
increased	The original price of $15 **increased** by $5.	$15+5$
more	3 coins and 8 **more**	$3+8$
more than	Josh has 10 points. Will has 5 **more than** Josh.	$10+5$
plus	8 baseballs **plus** 4 baseballs	$8+4$
sum	the **sum** of 3 and 5	$3+5$
total	the **total** of 10, 14, and 15	$10+14+15$

Words Indicating Subtraction	Example	Subtract
decreased	$16 **decreased** by $5	$16-5$
difference	the **difference** between 18 and 6	$18-6$
less	14 days **less** 5	$14-5$
less than	Jose completed 2 laps **less than** Mike's 9.	*$9-2$
left	Ray sold 15 out of 35 tickets. How many did he have **left**?	*$35-15$
lower than	This month's rainfall is 2 inches **lower than** last month's rainfall of 8 inches.	*$8-2$
minus	15 **minus** 6	$15-6$

* In subtraction word problems, you cannot always subtract the numbers in the order that they appear in the problem. Sometimes the first number should be subtracted from the last. You must read each problem carefully.

Words Indicating Multiplication	Example	Multiply
double	Her $1000 profit **doubled** in a month.	1000×2
half	**Half** of the $600 collected went to charity.	$\frac{1}{2}\times 600$
product	the **product** of 4 and 8	4×8
times	Li scored 3 **times** as many points as Ted who only scored 4.	3×4
triple	The bacteria **tripled** its original colony of 10,000 in just one day.	$3\times 10,000$
twice	Ron has 6 CDs. Tom has **twice** as many.	2×6

Words Indicating Division	Example	Divide
divide into, by, or among	The group of 70 **divided into** 10 teams	$70\div 10$ or $\frac{70}{10}$
quotient	the **quotient** of 30 and 6	$30\div 6$ or $\frac{30}{6}$

Match the phrase with the correct algebraic expression below. The answers will be used more than once.

A. $y - 2$

B. $2y$

C. $y + 2$

D. $\frac{y}{2}$

E. $2 - y$

1. 2 more than y
2. 2 divided into y
3. 2 less than y
4. twice y
5. the quotient of y and 2
6. y increased by 2
7. 2 less y
8. the product of 2 and y
9. y decreased by 2
10. y doubled
11. 2 minus y
12. the total of 2 and y

Now practice writing parts of algebraic expressions from the following word problems.

Example 3: the product of 3 and a number, t Answer: $3t$

13. 3 less than x
14. y divided among 10
15. the sum of t and 5
16. n minus 14
17. 5 times k
18. the total of z and 12
19. double the number b
20. x increased by 1
21. the quotient t and 4
22. half of a number y
23. bacteria culture, b, doubled
24. triple John's age y
25. a number, n, plus 4
26. quantity, t, less 6
27. 18 divided by a number, x
28. n feet lower than 10
29. 3 more than p
30. the product of 4 and m
31. a number, y, decreased by 20
32. 5 times as much as x

2.4 Setting Up Algebra Word Problems

So far, you have seen only the first part of algebra word problems. To complete an algebra problem, an equal sign must be added. The words "**is**" or "**are**" as well as "**equal(s)**" signal that you should add an equal sign.

Example 4: Double Jake's age, x, minus 4 is 22.

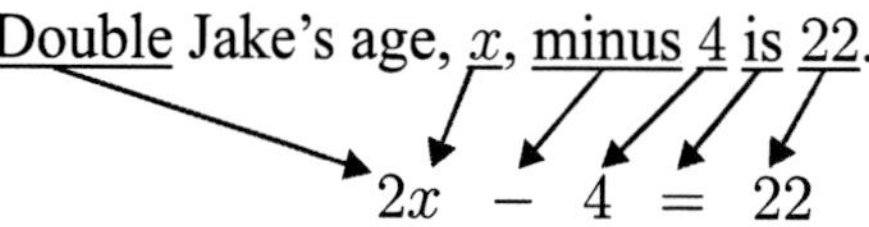

$$2x - 4 = 22$$

Translate the following word problems into algebra equations. DO NOT find the solutions to the problems yet.

1. Triple the original number, n, is $2,700$.
2. The product of a number, y, and 5 is equal to 15.
3. Four times the difference of a number, x, and 2 is 20.
4. The total, t, divided into 5 groups is 45.
5. The number of parts in inventory, p, minus 54 parts sold today is 320.
6. One-half an amount, x, added to \$50 is \$262
7. One hundred seeds divided by 5 rows equals n number of seeds per row.
8. A number, y, less than 50 is 82.
9. His base pay of \$200 increased by his commission, x, is \$500.
10. Seventeen more than half a number, h, is 35.
11. This month's sales of \$2,300 are double January's sales, x.
12. The quotient of a number, w, and 4 is 32.
13. Six less a number, d, is 12.
14. Four times the sum of a number, y, and 10 is 48.
15. We started with x number of students. When 5 moved away, we had 42 left.
16. A number, b, divided into 36 is 12.

2.5 Changing Algebra Word Problems to Algebraic Equations

Example 5: There are 3 people who have a total weight of 595 pounds. Sally weighs 20 pounds less than Jessie. Rafael weighs 15 pounds more than Jessie. How much does Jessie weigh?

Step 1: Notice everyone's weight is given in terms of Jessie. Sally weighs 20 pounds less than Jessie. Rafael weighs 15 pounds more than Jessie. First, we write everyone's weight in terms of Jessie, j.

$$\begin{aligned} \text{Jessie} &= j \\ \text{Sally} &= j - 20 \\ \text{Rafael} &= j + 15 \end{aligned}$$

Step 2: We know that all three together weigh 595 pounds. We write the sum of everyone's weight equal to 595.

$$j + j - 20 + j + 15 = 595$$

We will learn to solve these problems in the next chapter.

Change the following word problems to algebraic equations.

1. Fluffy, Spot, and Shampy have a combined age in dog years of 91. Spot is 14 years younger than Fluffy. Shampy is 6 years older than Fluffy. What is Fluffy's age, f, in dog years?

2. Jerry Marcosi puts 5% of the amount he makes per week into a retirement account, r. He is paid \$11.00 per hour and works 40 hours per week for a certain number of weeks, w. Write an equation to help him find out how much he puts into his retirement account.

3. A furniture store advertises a 40% off liquidation sale on all items. What would the sale price (p) be on a \$2530 dining room set?

4. Kyle Thornton buys an item which normally sells for a certain price, x. Today the item is selling for 25% off the regular price. A sales tax of 6% is added to the equation to find the final price, f.

5. Tamika Francois runs a floral shop. On Tuesday, Tamika sold a total of \$600 worth of flowers. The flowers cost her \$100, and she paid an employee to work 8 hours for a given wage, w. Write an equation to help Tamika find her profit, p, on Tuesday.

6. Sharice is a waitress at a local restaurant. She makes an hourly wage of \$3.50, plus she receives tips. On Monday, she works 6 hours and receives tip money, t. Write an equation showing what Sharice makes on Monday, y.

7. Jenelle buys x shares of stock in a company at \$34.50 per share. She later sells the shares at \$40.50 per share. Write an equation to show how much money, m, Jenelle has made.

2.6 Properties of Addition and Multiplication

The Associative, Commutative, Distributive, Identity, and Inverse properties of Addition and Multiplication are listed below by example as a quick refresher.

Property	Example
1. Associative Property of Addition	$(a+b)+c=a+(b+c)$
2. Associative Property of Multiplication	$(a\times b)\times c=a\times(b\times c)$
3. Commutative Property of Addition	$a+b=b+a$
4. Commutative Property of Multiplication	$a\times b=b\times a$
5. Distributive Property	$a\times(b+c)=(a\times b)+(a\times c)$
6. Identity Property of Addition	$0+a=a$
7. Identity Property of Multiplication	$1\times a=a$
8. Inverse Property of Addition	$a+(-a)=0$
9. Inverse Property of Multiplication	$a\times\frac{1}{a}=\frac{a}{a}=1, a\neq 0$

Write the number of the property listed above that describes each of the following statements.

1. $4+5=5+4$
2. $4+(2+8)=(4+2)+8$
3. $10\,(4+7)=(10)\,(4)+(10)\,(7)$
4. $(2\times 3)\times 4=2\times(3\times 4)$
5. $1\times 12=12$
6. $8\left(\frac{1}{8}\right)=1$
7. $1c=c$
8. $18+0=18$
9. $9+(-9)=0$
10. $p\times q=q\times p$
11. $t+0=t$
12. $x\,(y+z)=xy+xz$
13. $(m)\,(n\cdot p)=(m\cdot n)\,(p)$
14. $-y+y=0$

Chapter 2 Review

Solve the following problems using $x = 2$.

1. $3x + 4 =$
2. $\frac{6x}{4} =$
3. $x^2 - 5 =$
4. $\frac{x^3 + 8}{2} =$
5. $12 - 3x =$
6. $x - 5 =$
7. $-5x + 4 =$
8. $9 - x =$
9. $2x + 2 =$

Solve the following problems. Let $w = -1$, $y = 3$, $z = 5$.

10. $5w - y =$
11. $wyz + 2 =$
12. $z - 2w =$
13. $\frac{3z + 5}{wz} =$
14. $\frac{6w}{y} + \frac{z}{w} =$
15. $25 - 2yz =$
16. $-2y + 3$
17. $4w - (yw) =$
18. $7y - 5z =$

For the following questions, write an equation to match each problem.

19. Calista earns \$450 per week for a 40-hour work week plus \$16.83 per hour for each hour of overtime after 40 hours. Write an equation that would be used to determine her weekly wages where w is her wages and v is the number of overtime hours worked.

20. Daniel purchased a 1-year CD, c, from a bank. He bought it at an annual interest rate of 6%. After 1 year, Daniel cashes in the CD. What is the total amount it is worth?

21. Omar is a salesman. He earns an hourly wage of \$8.00 per hour, plus he receives a commission of 7% on the sales he makes. Write an equation which would be used to determine his weekly salary, w, where x is the number of hours worked, and y is the amount of sales for the week.

22. Tom earns \$500 per week before taxes are taken out. His employer takes out a total of 33% for state, federal, and Social Security taxes. Which expression below will help Tom figure his net pay?
    (A) $500 - 0.33$
    (B) $500 \div 0.33$
    (C) $500 + 0.33\,(500)$
    (D) $500 - 0.33\,(500)$

23. Rosa has to pay the first $100 of her medical expenses each year before she qualifies for her insurance company to begin paying. After paying the $100 "deductible," her insurance company will pay 80% of her medical expenses. This year, her total medical expenses came to $960.00. Which expression below shows how much her insurance company will pay?

(A) $0.80\,(960 - 100)$
(B) $100 + (960 \div 0.80)$
(C) $960\,(100 - 0.80)$
(D) $0.80\,(960 + 100)$

24. A plumber charges $45 per hour plus a $25.00 service charge. If a represents his total charges in dollars and b represents the number of hours worked, which formula below could the plumber use to calculate his total charges?

(A) $a = 45 + 25b$
(B) $a = 45 + 25 + b$
(C) $a = 45b + 25$
(D) $a = (45)\,(25) + b$

25. In 2004, Bell Computers informed its sales force to expect a 2.6% price increase on all computer equipment in the year 2005. A certain sales representative wanted to see how much the increase would be on a computer, c, that sold for $2200 in 2004. Which expression below will help him find the cost of the computer in the year 2005?

(A) $0.26\,(2200)$
(B) $2200 - 0.026\,(2200)$
(C) $2200 + 0.026\,(2200)$
(D) $0.026\,(2200) - 2200$

26. Juan sells a boat that he bought 5 years ago. He sells it for 60% less than he originally paid for it. If the original cost is b, write an expression that shows how much he sells the boat for.

27. Toshi is going to get a 7% raise after he works at his job for 1 year. If s represents his starting salary, write an expression that shows how much he will make after his raise.

Chapter 3
Solving Multi-Step Equations and Inequalities

This chapter covers the following North Carolina mathematics standards for Algebra I:

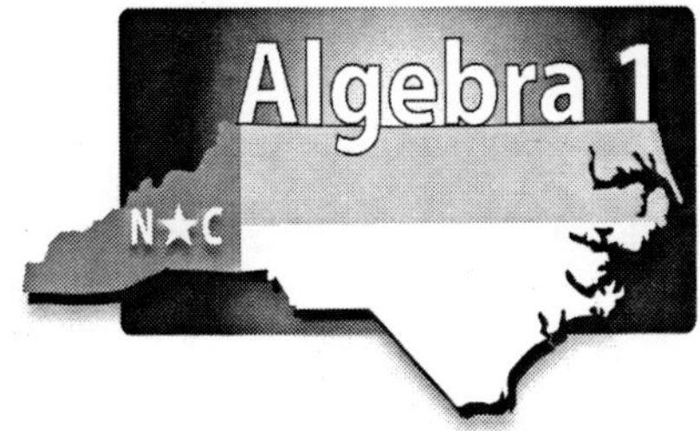

Competency Goal	Objectives
Number and Operations	1.01 1.02
Algebra	4.01a, b

3.1 Two-Step Algebra Problems

In the following two-step algebra problems, **additions** and **subtractions** are performed first and then **multiplication** and **division**.

Example 1: $-4x + 7 = 31$

Step 1: Subtract 7 from both sides.

$$\begin{array}{rr} -4x + 7 & = 31 \\ -7 & -7 \\ \hline -4x & = 24 \end{array}$$

Step 2: Divide both sides by -4.

$$\frac{\cancel{-4}x}{\cancel{-4}} = \frac{24}{-4} \quad \text{so} \quad x = \frac{24}{-4} = -6$$

Example 2: $-8 - y = 12$

Step 1: Add 8 to both sides.

$$\begin{array}{rr} -8 - y & = 12 \\ +8 & +8 \\ \hline -y & = 20 \end{array}$$

Step 2: To finish solving a problem with a negative sign in front of the variable, multiply both sides by -1. The variable needs to be positive in the answer.

$$(-1)(-y) = (-1)(20) \quad \text{so} \quad y = -20$$

Solve the two-step algebra problems below.

1. $6x - 4 = -34$
2. $5y - 3 = 32$
3. $8 - t = 1$
4. $10p - 6 = -36$
5. $11 - 9m = -70$
6. $4x - 12 = 24$
7. $3x - 17 = -41$
8. $9d - 5 = 49$
9. $10h + 8 = 78$
10. $-6b - 8 = 10$
11. $-g - 24 = -17$
12. $-7k - 12 = 30$
13. $9 - 5r = 64$
14. $6y - 14 = 34$
15. $12f + 15 = 51$
16. $21t + 17 = 80$
17. $20y + 9 = 149$
18. $15p - 27 = 33$
19. $22h + 9 = 97$
20. $-5 + 36w = 175$

3.2 Two-Step Algebra Problems with Fractions

An algebra problem may contain a fraction. Study the following example to understand how to solve algebra problems that contain a fraction.

Example 3: $\frac{x}{2} + 4 = 3$

Step 1:
$$\begin{array}{rr} \frac{x}{2} + 4 & = 3 \\ -4 & -4 \\ \hline \frac{x}{2} & = -1 \end{array}$$
Subtract 4 from both sides.

Step 2: $\frac{x}{2} = -1$ Multiply both sides by 2 to eliminate the fraction.

$\frac{x}{\cancel{2}} \times \cancel{2} = -1 \times 2$ so $x = -2$

Simplify the following algebra problems.

1. $4 + \frac{y}{3} = 7$
2. $\frac{a}{2} + 5 = 12$
3. $\frac{w}{5} - 3 = 6$
4. $\frac{x}{9} - 9 = -5$
5. $\frac{b}{6} + 2 = -4$
6. $7 + \frac{z}{2} = -13$
7. $\frac{x}{2} - 7 = 3$
8. $\frac{c}{5} + 6 = -2$
9. $3 + \frac{x}{11} = 7$
10. $16 + \frac{m}{6} = 14$
11. $\frac{p}{3} + 5 = -2$
12. $\frac{t}{8} + 9 = 3$
13. $\frac{v}{7} - 8 = -1$
14. $5 + \frac{h}{10} = 8$
15. $\frac{k}{7} - 9 = 1$
16. $\frac{y}{4} + 13 = 8$
17. $15 + \frac{z}{14} = 13$
18. $\frac{b}{6} - 9 = -14$
19. $\frac{d}{3} + 7 = 12$
20. $10 + \frac{b}{6} = 4$
21. $2 + \frac{p}{4} = -6$
22. $\frac{t}{7} - 9 = -5$
23. $\frac{a}{10} - 1 = 3$
24. $\frac{a}{8} + 16 = 9$

3.3 More Two-Step Algebra Problems with Fractions

Study the following example to understand how to solve algebra problems that contain a different type of fraction.

Example 4: $\dfrac{x+2}{4} = 3$

In this example, "$x + 2$" is divided by 4, and not just the x or the 2.

Step 1: $\dfrac{x+2}{\not{4}} \times \not{4} = 3 \times 4$
$x + 2 = 12$

First multiply both sides by 4 to eliminate the fraction.

Step 2:
$$\begin{array}{rr} x+2 & =12 \\ -2 & -2 \\ \hline x & =10 \end{array}$$

Next, subtract 2 from both sides.

Solve the following problems.

1. $\dfrac{x+1}{5} = 4$
2. $\dfrac{z-9}{2} = 7$
3. $\dfrac{b-4}{4} = -5$
4. $\dfrac{y-9}{3} = 7$
5. $\dfrac{d-10}{-2} = 12$
6. $\dfrac{w-10}{-8} = -4$
7. $\dfrac{x-1}{-2} = -5$
8. $\dfrac{c+40}{-5} = -7$
9. $\dfrac{13+h}{2} = 12$
10. $\dfrac{k-10}{3} = 9$
11. $\dfrac{a+11}{-4} = 4$
12. $\dfrac{x-20}{7} = 6$
13. $\dfrac{t+2}{6} = -5$
14. $\dfrac{b+1}{-7} = 2$
15. $\dfrac{f-9}{3} = 8$
16. $\dfrac{4+w}{6} = -6$
17. $\dfrac{3+t}{3} = 10$
18. $\dfrac{x+5}{5} = -3$
19. $\dfrac{g+3}{2} = 11$
20. $\dfrac{k+1}{-6} = 5$
21. $\dfrac{y-14}{2} = -8$
22. $\dfrac{z-4}{-2} = 13$
23. $\dfrac{w+2}{15} = -1$
24. $\dfrac{3+h}{3} = 6$

3.4 Rationalizing the Denominator

As we have seen, algebra problems can contain fractions. The problems can contain many different types of fractions. In this section, the algebra problems will have a fraction that contains a variable as the denominator. In other words, the bottom number of a fraction can be a variable in algebra problems.

Example 5: Solve $\frac{1}{x} + 5 = 7$ for x.

Step 1: The first thing to do is get all of the constants on one side of the equation. Do this by subtracting both sides by 5.

$\frac{1}{x} + 5 - 5 = 7 - 5$

$\frac{1}{x} + 0 = 2$

$\frac{1}{x} = 2$

Step 2: Next, multiply both sides of the equation by x to get x out the denominator of the fraction.

$\frac{1}{x} \times x = 2 \times x$

$1 = 2x$

Step 3: Last, divide both sides by 2 to get x on one side of the equation by itself.

$\frac{1}{2} = \frac{2x}{2}$

$\frac{1}{2} = x$ or $x = \frac{1}{2}$

Solve the following problems.

1. $\frac{4}{x} - 3 = 1$
2. $3 + \frac{2}{t} = 6$
3. $\frac{5}{p} + 5 = -10$
4. $\frac{12}{x} - 1 = 2$
5. $\frac{1}{2} = 2 + \frac{6}{n}$
6. $\frac{1}{x} - 15 = -9$
7. $23 - \frac{9}{f} = 5$
8. $\frac{-2}{a} + 7 = -3$
9. $-1 = 1 + \frac{8}{x}$
10. $12 = \frac{-18}{w} + 6$
11. $17 - \frac{2}{x} = 25$
12. $\frac{100}{z} - 13 = 7$
13. $25 - \frac{18}{t} = -11$
14. $\frac{1}{d} + 7 = 6$
15. $39 - \frac{7}{k} = -10$

3.5 Combining Like Terms

In an algebra problem, separate **terms** by + and − signs. The expression $5x - 4 - 3x + 7$ has 4 terms: $5x$, 4, $3x$, and 7. Terms having the same variable can be combined (added or subtracted) to simplify the expression. $5x - 4 - 3x + 7$ simplifies to $2x + 3$.

$$5x - 3x \quad - 4 + 7 \ = 2x + 3$$

Simplify the following expressions.

1. $7x + 12x$
2. $8y - 5y + 8$
3. $4 - 2x + 9$
4. $11a - 16 - a$
5. $9w + 3w + 3$
6. $-5x + x + 2x$
7. $w - 15 + 9w$
8. $21 - 10t + 9 - 2t$
9. $-3 + x - 4x + 9$
10. $7b + 12 + 4b$
11. $4h - h + 2 - 5$
12. $-6k + 10 - 4k$
13. $2a + 12a - 5 + a$
14. $5 + 9c - 10$
15. $-d + 1 + 2d - 4$
16. $-8 + 4h + 1 - h$
17. $12x - 4x + 7$
18. $10 + 3z + z - 5$
19. $14 + 3y - y - 2$
20. $11p - 4 + p$
21. $11m + 2 - m + 1$

3.6 Solving Equations with Like Terms

When an equation has two or more like terms on the same side of the equation, combining like terms is the **first** step in solving the equation.

Example 6: $7x + 2x - 7 = 21 + 8$

Step 1: Combine like terms on both sides of the equation.

Step 2: Solve the two-step algebra problem as explained previously.

$$\begin{aligned} 7x + 2x - 7 &= 21 + 8 \\ 9x - 7 &= 29 \\ +7 \quad & \quad +7 \\ 9x \div 9 &= 36 \div 9 \\ x &= 4 \end{aligned}$$

Solve the equations below combining like terms first.

1. $3w - 2w + 4 = 6$
2. $7x + 3 + x = 16 + 3$
3. $5 - 6y + 9y = -15 + 5$
4. $-14 + 7a + 2a = -5$
5. $-2t + 4t - 7 = 9$
6. $9d + d - 3d = 14$
7. $-6c - 4 - 5c = 10 + 8$
8. $15m - 9 - 6m = 9$
9. $-4 - 3x - x = -16$
10. $9 - 12p + 5p = 14 + 2$
11. $10y + 4 - 7y = -17$
12. $-8a - 15 - 4a = 9$

If the equation has like terms on both sides of the equation, you must get all of the terms with a **variable** on one side of the equation and all of the **integers** on the other side of the equation.

Example 7: $3x + 2 = 6x - 1$

Step 1: Subtract $6x$ from both sides to move all the **variables** to the left side.

Step 2: Subtract 2 from both sides to move all the **integers** to the right side.

Step 3: Divide by -3 to solve for x.

$$\begin{aligned} 3x + 2 &= 6x - 1 \\ -6x \quad &\quad\; -6x \\ -3x + 2 &= -1 \\ -2 &\quad\; -2 \\ \hline \frac{-3x}{-3} &= \frac{-3}{-3} \\ x &= 1 \end{aligned}$$

Solve the following problems.

1. $3a + 1 = a + 9$
2. $2d - 12 = d + 3$
3. $5x + 6 = 14 - 3x$
4. $15 - 4y = 2y - 3$
5. $9w - 7 = 12w - 13$
6. $10b + 19 = 4b - 5$
7. $-7m + 9 = 29 - 2m$
8. $5x - 26 = 13x - 2$
9. $19 - p = 3p - 9$
10. $-7p - 14 = -2p + 11$
11. $16y + 12 = 9y + 33$
12. $13 - 11w = 3 - w$
13. $-17b + 23 = -4 - 8b$
14. $k + 5 = 20 - 2k$
15. $12 + m = 4m + 21$
16. $7p - 30 = p + 6$
17. $19 - 13z = 9 - 12z$
18. $8y - 2 = 4y + 22$
19. $5 + 16w = 6w - 45$
20. $-27 - 7x = 2x + 18$
21. $-12x + 14 = 8x - 46$
22. $27 - 11h = 5 - 9h$
23. $5t + 36 = -6 - 2t$
24. $17y + 42 = 10y + 7$
25. $22x - 24 = 14x - 8$
26. $p - 1 = 4p + 17$
27. $4d + 14 = 3d - 1$
28. $7w - 5 = 8w + 12$
29. $-3y - 2 = 9y + 22$
30. $17 - 9m = m - 23$

3.7 Removing Parentheses

The distributive property is used to remove parentheses.

Example 8: $2(a+6)$

You multiply 2 by each term inside the parentheses. $2 \times a = 2a$ and $2 \times 6 = 12$. The 12 is a positive number so use a plus sign between the terms in the answer.

$2(a+6) = 2a + 12$

Example 9: $4(-5c+2)$

The first term inside the parentheses could be negative. Multiply in exactly the same way as the examples above. $4 \times (-5c) = -20c$ and $4 \times 2 = 8$

$4(-5c+2) = -20c + 8$

Remove the parentheses in the problems below and simplify the following expressions.

1. $7(n+6)$
2. $8(2g-5)$
3. $11(5z-2)$
4. $6(-y-4)$
5. $3(-3k+5)$
6. $4(d-8)$
7. $2(-4x+6)$
8. $7(4+6p)$
9. $5(-4w-8)$
10. $6(11x+2)$
11. $10(9-y)$
12. $9(c-9)$
13. $12(-3t+1)$
14. $3(4y+9)$
15. $8(b+3)$

The number in front of the parentheses can also be negative. Remove these parentheses the same way.

Example 10: $-2(b-4)$

First, multiply $-2 \times b = -2b$

Second, multiply $-2 \times 4 = -8$

Copy the two products. The second product is a positive number so put a plus sign between the terms in the answer.

$2(b-4) = -2b + 8$

Remove the parentheses in the following problems and simplify the following expressions.

16. $-7(x+2)$
17. $-5(4-y)$
18. $-4(2b-2)$
19. $-2(8c+6)$
20. $-5(-w-8)$
21. $-3(4x-2)$
22. $-2(-z+2)$
23. $-4(7p+7)$
24. $-9(t-6)$
25. $-10(2w+4)$
26. $-3(9-7p)$
27. $-9(-k-3)$
28. $-1(7b-9)$
29. $-6(-5t-2)$
30. $-7(-v+4)$

3.8 Multi-Step Algebra Problems

You can now use what you know about removing parentheses, combining like terms, and solving simple algebra problems to solve problems that involve three or more steps. Study the examples below to see how easy it is to solve multi-step problems.

Example 11: $3(x+6) = 5x - 2$

Step 1: Use the distributive property to remove parentheses. $3x + 18 = 5x - 2$

Step 2: Subtract $5x$ from each side to move the terms with variables to the left side of the equation.

$$\begin{array}{rr} 3x + 18 & = 5x - 2 \\ -5x & -5x \\ \hline -2x + 18 & = -2 \end{array}$$

Step 3: Subtract 18 from each side to move the integers to the right side of the equation.

$$\begin{array}{rr} -18 & -18 \\ \hline \frac{-2x}{-2} & = \frac{-20}{-2} \\ x & = 10 \end{array}$$

Step 4: Divide both sides by -2 to solve for x.

Example 12: $\dfrac{3(x-3)}{2} = 9$

Step 1: Use the distributive property to remove parentheses. $\dfrac{3x-9}{2} = 9$

Step 2: Multiply both sides by 2 to eliminate the fraction. $\dfrac{2(3x-9)}{2} = 2(9)$

Step 3: Add 9 to both sides, and combine like terms.

$$\begin{array}{rr} 3x - 9 & = 18 \\ +9 & +9 \\ \hline \frac{3x}{3} & = \frac{27}{3} \\ x & = 9 \end{array}$$

Step 4: Divide both sides by 3 to solve for x.

Solve the following multi-step algebra problems.

1. $2(y-3) = 4y + 6$
2. $\dfrac{2(a+4)}{2} = 12$
3. $\dfrac{10(x-2)}{5} = 14$
4. $\dfrac{12y-18}{6} = 4y + 3$
5. $2x + 3x = 30 - x$
6. $\dfrac{2a+1}{3} = a + 5$
7. $5(b-4) = 10b + 5$
8. $-8(y+4) = 10y + 4$

9. $\dfrac{x+4}{-3} = 6 - x$

10. $\dfrac{4(n+3)}{5} = n - 3$

11. $3(2x-5) = 8x - 9$

12. $7 - 10a = 9 - 9a$

13. $7 - 5x = 10 - (6x+7)$

14. $4(x-3) - x = x - 6$

15. $4a + 4 = 3a - 4$

16. $-3(x-4) + 5 = -2x - 2$

17. $5b - 11 = 13 - b$

18. $\dfrac{-4x+3}{2x} = \dfrac{7}{2x}$

19. $-(x+1) = -2(5-x)$

20. $4(2c+3) - 7 = 13$

21. $6 - 3a = 9 - 2(2a+5)$

22. $-5x + 9 = -3x + 11$

23. $3y + 2 - 2y - 5 = 4y + 3$

24. $3y - 10 = 4 - 4y$

25. $-(a+3) = -2(2a+1) - 7$

26. $5m - 2(m+1) = m - 10$

27. $\dfrac{1}{2}(b-2) = 5$

28. $-3(b-4) = -2b$

29. $4x + 12 = -2(x+3)$

30. $\dfrac{7x+4}{3} = 2x - 1$

31. $9x - 5 = 8x - 7$

32. $7x - 5 = 4x + 10$

33. $\dfrac{4x+8}{2} = 6$

34. $2(c+4) + 8 = 10$

35. $y - (y+3) = y + 6$

36. $4 + x - 2(x-6) = 8$

3.9 Solving Radical Equations

Some multi-step equations contain radicals. An example of a radical is a square root, $\sqrt{}$.

Example 13: Solve the following equation for x. $\sqrt{4x-3}+2=5$

Step 1: The first step is to get the constants that are not under the radical on one side. Subtract 2 from both sides of the equation.
$\sqrt{4x-3}+2-2=5-2$
$\sqrt{4x-3}+0=3$
$\sqrt{4x-3}=3$

Step 2: Next, you must get rid of the radical sign by squaring both sides of the equation.
$\left(\sqrt{4x-3}\right)^2=(3)^2$
$4x-3=9$

Step 3: Add 3 to both sides of the equation to get the constants on just one side of the equation.
$4x-3+3=9+3$
$4x+0=12$
$4x=12$

Step 4: Last, get x on one side of the equation by itself by dividing both sides by 4.
$\frac{4x}{4}=\frac{12}{4}$
$x=3$

Solve the following equations.

1. $\sqrt{x+3}-13=-8$
2. $3+\sqrt{7t-3}=5$
3. $\sqrt{3q+12}-4=5$
4. $\sqrt{11f+3}+2=8$
5. $5=\sqrt{6g-5}+(-2)$
6. $-2=\sqrt{x-3}$
7. $\sqrt{-8t}-3=1$
8. $\sqrt{-d+1}-9=-6$
9. $10-\sqrt{8x+2}=9$
10. $\sqrt{15y+4}+4=-4$
11. $\sqrt{r+14}-1=9$
12. $3-\sqrt{2q-1}=6$
13. $\sqrt{5t+16}+4=13$
14. $17=\sqrt{23-f}+15$
15. $19-\sqrt{7x-5}=22$

3.10 Multi-Step Inequalities

Remember that adding and subtracting with inequalities follow the same rules as equations. When you multiply or divide both sides of an inequality by the same positive number, the rules are also the same as for equations. However, when you multiply or divide both sides of an inequality by a **negative** number, you must **reverse** the inequality symbol.

Example 14: $-x > 4$

$(-1)(-x) < (-1)(4)$

$x < -4$

Example 15: $-4x < 2$

$\frac{-4x}{-4} > \frac{2}{-4}$

$x > -\frac{1}{2}$

Reverse the symbol when you multiply or divide by a negative number.

When solving multi-step inequalities, first add and subtract to isolate the term with the variable. Then multiply and divide.

Example 16: $2x - 8 > 4x + 1$

Step 1: Add 8 to both sides.

$2x - 8 + 8 > 4x + 1 + 8$

$2x > 4x + 9$

Step 2: Subtract $4x$ from both sides.

$2x - 4x > 4x + 9 - 4x$

$-2x > 9$

Step 3: Divide by -2. Remember to change the direction of the inequality sign.

$\frac{-2x}{-2} < \frac{9}{-2}$

$x < -\frac{9}{2}$

Solve each of the following inequalities.

1. $8 - 3x \leq 7x - 2$

2. $3(2x - 5) \geq 8x - 5$

3. $\frac{1}{3}b - 2 > 5$

4. $7 + 3y > 2y - 5$

5. $3a + 5 < 2a - 6$

6. $3(a - 2) > -5a - 2(3 - a)$

7. $2x - 7 \geq 4(x - 3) + 3x$

8. $6x - 2 \leq 5x + 5$

9. $-\frac{x}{4} > 12$

10. $-\frac{2x}{3} \leq 6$

11. $3b + 5 < 2b - 8$

12. $4x - 5 \leq 7x + 13$

13. $4x + 5 \leq -2$

14. $2y - 5 > 7$

15. $4 + 2(3 - 2y) \leq 6y - 20$

16. $-4c + 6 \leq 8$

17. $-\frac{1}{2}x + 2 > 9$

18. $\frac{1}{4}y - 3 \leq 1$

19. $-3x + 4 > 5$

20. $\frac{y}{2} - 2 \geq 10$

21. $7 + 4c < -2$

22. $2 - \frac{a}{2} > 1$

23. $10 + 4b \leq -2$

24. $-\frac{1}{2}x + 3 > 4$

3.11 Solving Equations and Inequalities with Absolute Values

When solving equations and inequalities which involve variables placed in absolute values, remember that there will be two or more numbers that will work as correct answers. This is because the absolute value variable will signify both positive and negative numbers as answers.

Example 17: $5 + 3|k| = 8$ Solve as you would any equation.

Step 1: $3|k| = 3$ Subtract 5 from each side.

Step 2: $|k| = 1$ Divide by 3 on each side.

Step 3: $k = 1$ or $k = -1$ Because k is an absolute value, the answer can be 1 or -1

Example 18: $2|x| - 3 < 7$ Solve as you normally would an inequality.

Step 1: $2|x| < 10$ Add 3 to both sides.

Step 2: $|x| < 5$ Divide by 2 on each side.

Step 3: $x < 5$ or $x > -5$ or $-5 < x < 5$ Because x is an absolute value, the answer is a set of both positive and negative numbers.

Read each problem, and write the number or set of numbers which solves each equation or inequality.

1. $7 + 2|y| = 15$
2. $4|x| - 9 < 3$
3. $6|k| + 2 = 14$
4. $10 - 4|n| > -14$
5. $-3 = 5|z| + 12$
6. $-4 + 7|m| < 10$
7. $5|x| - 12 > 13$
8. $21|g| + 7 = 49$
9. $-9 + 6|x| = 15$
10. $12 - 6|w| > -12$
11. $31 > 13 + 9|r|$
12. $-30 = 21 - 3|t|$
13. $9|x| - 19 < 35$
14. $-13|c| + 21 \geq -31$
15. $5 - 11|k| < -17$
16. $-42 + 14|p| = 14$
17. $15 < 3|b| + 6$
18. $9 + 5|q| = 29$
19. $-14|y| - 38 < -45$
20. $36 = 4|s| + 20$
21. $20 \leq -60 + 8|e|$

3.12 More Solving Equations and Inequalities with Absolute Values

Now, look at the following examples in which numbers and variables are added or subtracted within the absolute value symbols (||).

Example 19: $|3x - 5| = 10$

Remember an equation with absolute value symbols has two solutions.

Step 1:

$3x - 5 = 10$
$3x - 5 + 5 = 10 + 5$
$\frac{\not{3}x}{\not{3}} = \frac{15}{3}$
$x = 5$

To find the first solution, remove the absolute value symbol and solve the equation.

Step 2:

$-(3x - 5) = 10$
$-3x + 5 = 10$
$-3x + 5 - 5 = 10 - 5$
$-3x = 5$
$x = -\frac{5}{3}$

To find the second solution, solve the equation for the negative of the expression in absolute value symbols.

Solutions: $x = \left\{5, -\frac{5}{3}\right\}$

Example 20: $|5z - 10| < 20$

Remove the absolute value symbols and solve the inequality.

Step 1:

$5z - 10 < 20$
$5z - 10 + 10 < 20 + 10$
$\frac{\not{5}z}{\not{5}} < \frac{30}{5}$
$z < 6$

Step 2:

$-(5z - 10) < 20$
$-5z + 10 < 20$
$-5z + 10 - 10 < 20 - 10$
$\frac{-\not{5}z}{\not{5}} < \frac{10}{5}$
$-z < 2$
$z > -2$

Next, solve the equation for the negative of the expression in the absolute value symbols.

Solution: $-2 < z < 6$

Example 21: $|4y + 7| - 5 > 18$

Step 1: $4y + 7 - 5 + 5 > 18 + 5$ Remove the absolute value symbols and solve the inequality.
$4y + 7 > 23$
$4y + 7 - 7 > 23 - 7$
$4y > 16$
$y > 4$

Step 2: $-(4y + 7) - 5 > 18$ Solve the equation for the negative of the expression in the absolute value symbols.
$-4y - 7 - 5 + 5 > 18 + 5$
$-4y - 7 + 7 > 23 + 7$
$-4y > 30$
$y < -7\frac{1}{2}$

Solutions: $y > 4$ or $y < -7\frac{1}{2}$

Solve the following equations and inequalities below.

1. $-4 + |2x + 4| = 14$
2. $|4b - 7| + 3 > 12$
3. $6 + |12e + 3| < 39$
4. $-15 + |8f - 14| > 35$
5. $|-9b + 13| - 12 = 10$
6. $-25 + |7b + 11| < 35$
7. $|7w + 2| - 60 > 30$
8. $63 + |3d - 12| = 21$
9. $|-23 + 8x| - 12 > +37$
10. $|61 + 20x| + 32 > 51$
11. $|4a + 13| + 31 = 50$
12. $4 + |4k - 32| < 51$
13. $8 + |4x + 3| = 21$
14. $|28 + 7v| - 28 < 77$
15. $|62p + 31| + 43 = 136$
16. $18 - |6v + 22| < 22$
17. $12 = 4 + |42 + 10m|$
18. $53 < 18 + |12e + 31|$
19. $38 > -39 + |7j + 14|$
20. $9 = |14 + 15u| + 7$
21. $11 - |2j + 50| > 45$
22. $|35 + 6i| - 3 = 14$
23. $|26 - 8r| - 9 > 41$
24. $|25 + 6z| - 21 = 28$
25. $12 < |2t + 6| - 14$
26. $50 > |9q - 10| + 6$
27. $12 + |8v - 18| > 26$
28. $-38 + |16i - 33| = 41$
29. $|-14 + 6p| - 9 < 7$
30. $28 > |25 - 5f| - 12$

Chapter 3 Review

Solve each of the following equations.

1. $4a - 8 = 28$
2. $5 + \frac{x}{8} = -4$
3. $-7 + 23w = 108$
4. $\frac{y-8}{6} = 7$
5. $c - 13 = 5$
6. $\frac{b+9}{12} = -3$

Solve.

7. $19 - 8d = d - 17$
8. $\frac{-3}{x} + 11 = -1$
9. $7w - 8w = -4w - 30$
10. $3 - \sqrt{6 - 2x} = 4$
11. $\frac{12}{f} - 7 = -5$
12. $6 + 16x = -2x - 12$
13. $\sqrt{w + 11} + 14 = 15$
14. $6 - \frac{1}{q} = 4$
15. $\sqrt{5k + 1} - 3 = 11$

Remove parentheses.

16. $3(-4x + 7)$
17. $11(2y + 5)$
18. $6(8 - 9b)$
19. $-8(-2 + 3a)$
20. $-2(5c - 3)$
21. $-5(7y - 1)$

Solve for the variable.

22. If $3x - y = 15$, then $y =$

23. If $7a + 2b = 1$, then $b =$

Solve each of the following equations and inequalities.

24. $\frac{-11c - 35}{4} = 4c - 2$
25. $5 + x - 3(x + 4) = -17$
26. $4(2x + 3) \geq 2x$
27. $7 - 3x \leq 6x - 2$
28. $\frac{5(n+4)}{3} = n - 8$
29. $-y > 14$
30. $2(3x - 1) \geq 3x - 7$
31. $3(x + 2) < 7x - 10$

Chapter 4
Ratios and Proportions

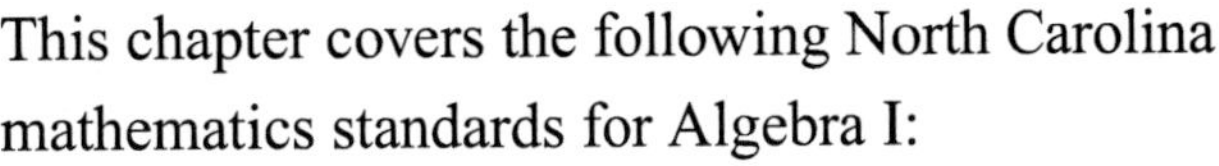
This chapter covers the following North Carolina mathematics standards for Algebra I:

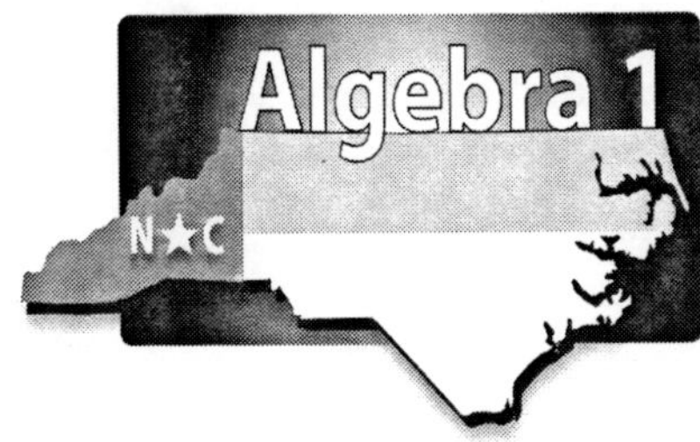

Competency Goal	Objectives
Number and Operations	1.01
	1.02
	1.03
Algebra	4.01

4.1 Ratio Problems

In some word problems, you may be asked to express answers as a **ratio**. Ratios can look like fractions. Numbers must be written in the order they are requested. In the following problem, 8 cups of sugar is mentioned before 6 cups of strawberries. But in the question part of the problem, you are asked for the ratio of STRAWBERRIES to SUGAR. The amount of strawberries IS THE FIRST WORD MENTIONED, so it must be the **top** number of the fraction. The amount of sugar, THE SECOND WORD MENTIONED, must be the **bottom** number of the fraction.

Example 1: The recipe for jam requires 8 cups of sugar for every 6 cups of strawberries. What is the ratio of strawberries to sugar in this recipe?

First number requested $\underline{6}$ $\frac{\text{cups strawberries}}{\text{cups sugar}}$
Second number requested 8

Answers may be reduced to lowest terms. $\frac{6}{8} = \frac{3}{4}$

This ratio could also be expressed as 3 : 4.

Practice writing ratios for the following word problems and reduce to lowest terms. DO NOT CHANGE ANSWERS TO MIXED NUMBERS. Ratios should be left in fraction form.

1. Out of the 248 seniors, 112 are boys. What is the ratio of boys to the total number of seniors?
2. It takes 7 cups of flour to make 2 loaves of bread. What is the ratio of cups of flour to loaves of bread?
3. A skyscraper that stands 620 feet tall casts a shadow that is 125 feet long. What is the ratio of the shadow to the height of the skyscraper?
4. Twenty boxes of paper weigh 520 pounds. What is the ratio of boxes to pounds?
5. The newborn weighs 8 pounds and is 22 inches long. What is the ratio of weight to length?
6. Jack pays $6.00 for 10 pounds of apples. What is the ratio of the price of apples to the pounds of apples?
7. Jordan spends $45 on groceries. Of that total, $23 is for steaks. What is the ratio of steak cost to the total grocery cost?
8. Madison's flower garden measures 8 feet long by 6 feet wide. What is the ratio of length to width?

4.2 Solving Proportions

Two **ratios (fractions)** that are **equal** to each other are called **proportions. For example,** $\frac{1}{4} = \frac{2}{8}$**. Read the following example to see how to find a number missing from a proportion.**

Example 2: $\frac{5}{15} = \frac{8}{x}$

Step 1: To find x, you first multiply the two numbers that are diagonal to each other.

$$\frac{5}{\{15\}} = \frac{\{8\}}{x}$$

$15 \times 8 = 120$

$5 \times x = 5x$

Therefore, $5x = 120$

Step 2: Then divide the product (120) by the coefficient of the variable (5).

$120 \div 5 = 24$

Therefore, $\frac{5}{15} = \frac{8}{24}$ **and** $x = 24$.

Practice finding the number missing from the following proportions. First, multiply the two numbers that are diagonal from each other. Then divide by the other number.

1. $\frac{2}{5} = \frac{6}{x}$
2. $\frac{9}{3} = \frac{x}{5}$
3. $\frac{x}{12} = \frac{3}{4}$
4. $\frac{7}{x} = \frac{3}{9}$
5. $\frac{12}{x} = \frac{2}{5}$
6. $\frac{12}{x} = \frac{4}{3}$
7. $\frac{27}{3} = \frac{x}{2}$
8. $\frac{1}{x} = \frac{3}{12}$
9. $\frac{15}{2} = \frac{x}{4}$
10. $\frac{7}{14} = \frac{x}{6}$
11. $\frac{5}{6} = \frac{10}{x}$
12. $\frac{4}{x} = \frac{3}{6}$
13. $\frac{x}{5} = \frac{9}{15}$
14. $\frac{9}{18} = \frac{x}{2}$
15. $\frac{5}{7} = \frac{35}{x}$
16. $\frac{x}{2} = \frac{8}{4}$
17. $\frac{15}{20} = \frac{x}{8}$
18. $\frac{x}{40} = \frac{5}{100}$

4.3 Ratio and Proportion Word Problems

Example 3: A stick one meter long is held perpendicular to the ground and casts a shadow 0.4 meters long. At the same time, an electrical tower casts a shadow 112 meters long. Use ratio and proportion to find the height of the tower.

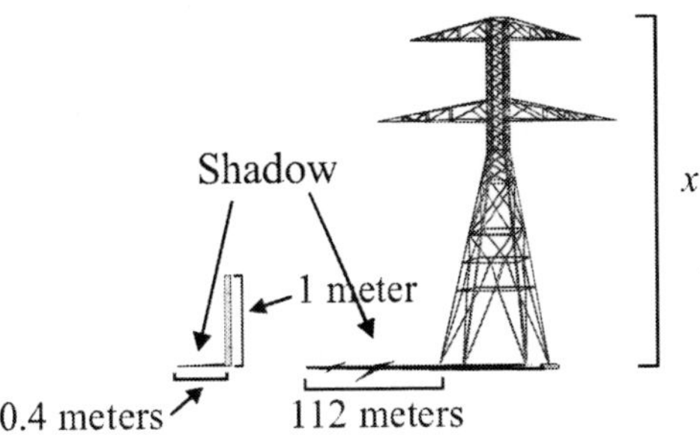

Step 1: Set up a proportion using the numbers in the problem. Put the shadow lengths on one side of the equation and put the heights on the other side. The 1 meter height is paired with the 0.4 meter length, so let them both be top numbers. Let the unknown height be x.

shadow length object height

$$\frac{0.4}{112} = \frac{1}{x}$$

Step 2: Solve the proportion as you did on page 54.

$112 \times 1 = 112$ $\qquad$ $112 \div 0.4 = 280$

Answer: The tower height is 280 meters.

Use ratio and proportion to solve the following problems.

1. Rudolph can mow a lawn that measures 1000 square feet in 2 hours. At that rate, how long would it take him to mow a lawn 3500 square feet?
2. Faye wants to know how tall her school building is. On a sunny day, she measures the shadow of the building to be 6 feet. At the same time she measures the shadow cast by a 5 foot statue to be 2 feet. How tall is her school building?
3. Out of every 5 students surveyed, 2 listen to country music. At that rate, how many students in a school of 800 listen to country music?
4. Butterfly, a Labrador Retriever, has a litter of 8 puppies. Four are black. At that rate, how many puppies in a litter of 10 would be black?
5. According to the instructions on a bag of fertilizer, 5 pounds of fertilizer are needed for every 100 square feet of lawn. How many square feet will a 25-pound bag cover?
6. A race car can travel 2 laps in 5 minutes. At this rate, how long will it take the race car to complete 100 laps ?
7. If it takes 7 cups of flour to make 4 loaves of bread, how many loaves of bread can you make from 35 cups of flour?
8. If 3 pounds of jelly beans cost $6.30, how much would 2 pounds cost?
9. For the first 4 home football games, the concession stand sold a total of 600 hotdogs. If that ratio stays constant, how many hotdogs will sell for all 10 home games?

4.4 Direct and Indirect Variation

The graphs shown below represent functions where x varies with y directly or indirectly. In direct variation, when y increases, the x increases, and when y decreases, x decreases. In indirect variation, also called inverse variation, when y increases, x decreases, and when y decreases, x increases.

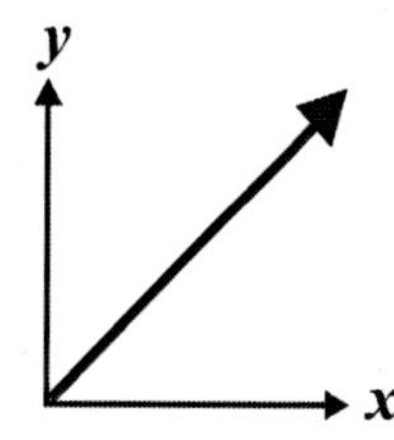

Direct Variation

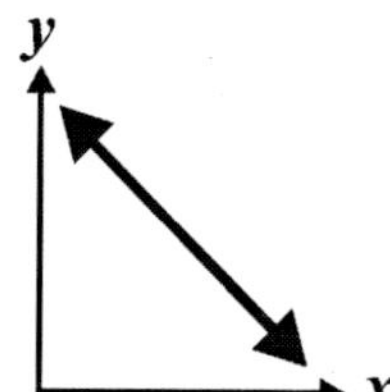

Indirect Variation

Example 4: Direct and indirect variation can be demonstrated with function tables.

Table 1

x	y
0	3
1	4
2	7
3	12
4	19

Table 2

x	y
0	20
1	18
2	16
3	14
4	12

Notice in Table 1, as x increases, y increases also. This means that function Table 1 represents a direct variation between x and y. On the other hand, Table 2 shows a decrease in y when x increases. This means that function Table 2 represents an indirect variation between x and y.

Direct variation occurs in a function when y varies directly, or in the same way, as x varies. The two values vary by a proportional factor, k. The variation is treated just like a proportion.

Example 5: If y varies directly with x, and $y = 18$ when $x = 12$, what is the value of y when $x = 6$?

Step 1: Set up the values in a proportion like you did in the previous two sections. Be sure to put the correct corresponding x and y values on the same line within the fractions.

x values y values

$$\frac{12}{6} = \frac{18}{y}$$

Step 2: Solve for y by multiplying the diagonals together and setting them equal to one another.

$12 \times y = 6 \times 18$

$\frac{12y}{12} = \frac{108}{12}$ Divide both sides by 12.

$y = 9$

For an **indirect variation**, y varies inversely with, or opposite of, x. With indirect variation, when x increases, y decreases, and when x decreases, y increases.

Example 6: In a function, y varies inversely as x varies. If $y = 18$ when $x = 12$, what is the value of y when $x = 6$?

Step 1: Set up the problem as a regular proportion problem, like Example 5.

x values		y values
$\frac{12}{6}$	$=$	$\frac{18}{y}$

Step 2: Now switch the numerator and denominator of the y values. This allows the x and y values to vary indirectly with each other. (Turn the fraction upside down.)

x values		y values
$\frac{12}{6}$	$=$	$\frac{y}{18}$

Step 3: Solve for y by multiplying the diagonals together.

$12 \times 18 = 6 \times y$

$216 = 6y$ Divide both sides by 6.

$36 = y$

Note: In an indirect variation problem, the reciprocal may be used for either side of the equation. In this case the x values or the y values could be switched to get the same value for y.

Example 7: It takes 45 minutes for 2 copiers to finish a printing job. If 5 copiers work together to print a job, how long would it take to finish?

Step 1: It will take less time to finish a job if more copiers work together. As the number of copiers increases, the number of minutes to complete the job decreases. Therefore, this is an indirect variation problem.

Step 2: Let y represent the number of minutes to complete the job. Let x represent the number of copiers. The old values are $y = 45$ minutes and $x = 2$ copiers, and the new value of x is 5 copiers. We are looking for the new y value.

Step 3: Set up the problem as a regular proportion problem.

x values		y values
$\frac{2}{5}$	$=$	$\frac{45}{y}$

Step 4: Now switch the numerator and denominator of the y values. This allows the x and y values to vary indirectly with each other. (Turn the fraction upside down.)

x values		y values
$\frac{2}{5}$	$=$	$\frac{y}{45}$

x values represent number of copiers
y values represent number of minutes

Step 5: Solve for y by multiplying the diagonals together.

$2 \times 45 = 5 \times y$

$90 = 5y$ Divide both sides by 5.

$18 = y$

It will take 5 copiers only 18 minutes to complete the printing job.

Solve these direct variation problems.

1. If $y = 6$ and $x = 3$, what is the value of y when $x = 5$?
2. If $y = 10$ and $x = 5$, what is the value of y when $x = 4$?
3. If $y = 6$ and $x = 2$, what is the value of y when $x = 7$?
4. If $y = 8$ and $x = 4$, what is the value of y when $x = 6$?
5. If $y = 15$ and $x = 3$, what is the value of y when $x = 5$?

Solve these indirect variation problems.

6. If $y = 6$ and $x = 4$, what is the value of y when $x = 8$?
7. If $y = 12$ and $x = 6$, what is the value of y when $x = 8$?
8. If $y = 9$ and $x = 6$, what is the value of y when $x = 3$?
9. If $y = 6$ and $x = 5$, what is the value of y when $x = 3$?
10. If $y = 3$ and $x = 12$, what is the value of y when $x = 9$?

Solve the following indirect word problems.

11. It takes an average person 60 minutes to type 8 pages on the computer. If three average typists work together to type up an 8-page paper, how long will it take them?

12. Sandra has $40 saved from her allowance this month to rent movies and buy books. If she buys 6 books, she will only be able to rent 2 movies. How many movies will she be able to rent if she only buys 4 books?

Solve the following direct and indirect word problems.

13. At the local grocery store, 2 pineapples cost $2.78. How much do 5 pineapples cost?

14. When Samuel rides his bike at a speed of 22 mph, it takes him 30 minutes to get home from school. Today he needs to be home in 25 minutes. How fast must he ride his bike to get home in time?

15. Jim must help his father carry bags of soil to the backyard. There are 45 bags of soil. Normally this would take Jim 30 minutes to do by himself, but today three of his friends stopped by and offered to help. How long will it take the four boys to carry all of those bags to the backyard?

16. It normally takes Jessica 45 minutes to get her friend's house 20 miles away. Tomorrow she is meeting her friend at the mall, which is 28 miles away from her house. If she travels at the same rate she normally travels, how long will it take her to get to the mall?

Chapter 4 Review

Solve the following proportions and ratios.

1. $\frac{8}{x} = \frac{1}{2}$ 2. $\frac{2}{5} = \frac{x}{10}$ 3. $\frac{x}{6} = \frac{3}{9}$ 4. $\frac{4}{9} = \frac{8}{x}$

5. Out of 100 coins, 45 are in mint condition. What is the ratio of mint condition coins to the total number of coins?

6. The ratio of boys to girls in the seventh grade is 6 : 5. If there are 135 girls in the class, how many boys are there?

7. Twenty out of the total 235 seniors graduate with honors. What is the ratio of seniors graduating with honors to the total number of seniors?

8. Aunt Bess uses 3 cups of oatmeal to bake 6 dozen oatmeal cookies. How many cups of oatmeal would she need to bake 15 dozen cookies?

9. If $y = 12$ and $x = 6$, using indirect variation, what is the value of y when $x = 20$?

10. If $y = 10$ and $x = 5$, what is the value of y when $x = 4$? Use direct variation to solve.

11. If $y = 5$ and $x = 2$, what is the value of y when $x = 12$? Use indirect variation to solve.

12. If $y = 42$ and $x = 6$, what is the value of y when $x = 12$? Use direct variation to solve.

13. Sandra waters 30 plants in 10 minutes. At this same rate, how many plants can she water in 25 minutes?

14. Every week, five friends buy the newest video game by putting their allowances together. All the new video games cost \$50, so each person gives \$10 towards the game. This week one of the five friends is grounded and doesn't receive an allowance, but the remaining four friends still want to buy a video game. How much will each friend have to contribute to buy the game?

Chapter 5
Algebra Word Problems

This chapter covers the following North Carolina mathematics standards for Algebra I:

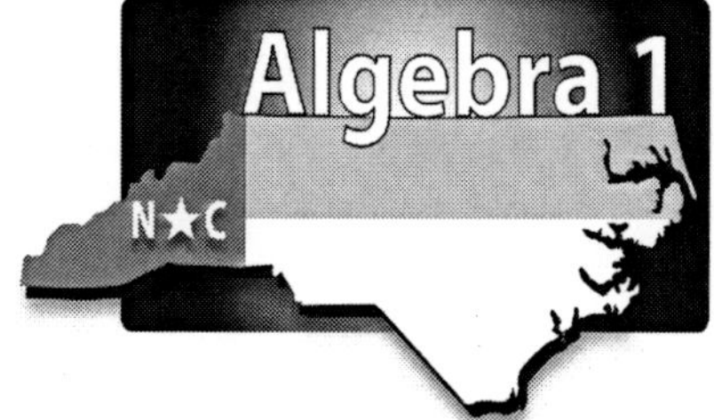

Competency Goal	Objectives
Number and Operations	1.01 1.02
Algebra	4.01

An equation states that two mathematical expressions are equal. In working with word problems, the words that mean equal are **equals, is, was, is equal to, amounts to,** and other expressions with the same meaning. To translate a word problem into an algebraic equation, use a variable to represent the unknown or unknowns you are looking for.

Example 1: Four more than twice a number is two less than three times the number.

Step 1: **Translation:** $4 + 2n = 3n - 2$

Step 2: **Now Solve:**

$$\begin{array}{rcl} 4 + 2n & = & 3n - 2 \\ -2n & & -2n \\ \hline 4 & = & n - 2 \\ +2 & & +2 \\ \hline 6 & = & n \end{array}$$

The number is 6.

Substitute the number back into the original equation to check.

Translate the following word problems into equations and solve.

1. Seven less than twice a number is eleven. Find the number.
2. Four more than three times a number is one less than four times the number. What is the number?
3. The sum of three times a number and the number is 24. What is the number?
4. Negative 16 is the sum of five and a number. Find the number.
5. Negative 20 is equal to ten minus the product of six and a number. What is the number?
6. Two less than twice a number equals the number plus 15. What is the number?
7. The difference between three times a number and 21 is three. What is the number?
8. Eighteen is fifteen less than the product of a number and three. What is the number?
9. Six more than twice a number is four times the difference between three and the number. What is the number?
10. Four less than twice a number is five times the sum of one and the number. What is the number?

5.1 Geometry Word Problems

The perimeter of a geometric figure is the distance around the outside of the figure.

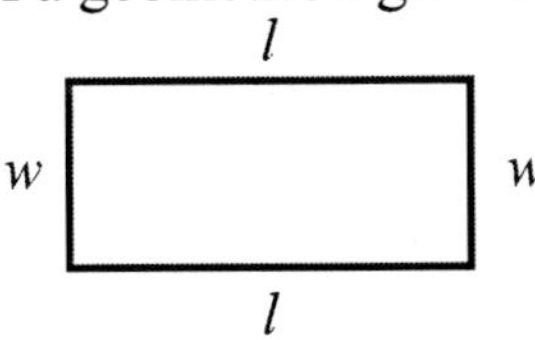

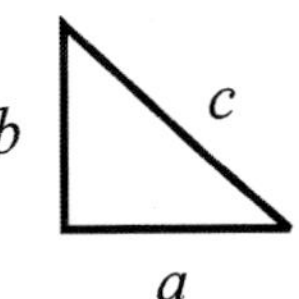

perimeter = $2l + 2w$ perimeter = $a + b + c$

Example 2: The perimeter of a rectangle is 44 feet. The length of the rectangle is 6 feet more than the width. What is the measure of the width?

Step 1: width $= w$ length $= 6 + w$

Step 2: Use the equation for the perimeter of a rectangle as follows:
$2l + 2w$ = perimeter of a rectangle
$2(w + 6) + 2w = 44$

Step 3: Solve for w.

Solution: width $= 8$ feet

Example 3: The perimeter of a triangle is 26 feet. The second side is twice as long as the first. The third side is 1 foot longer than the second side. What are the lengths of the 3 sides?

Step 1: Let $x =$ first side $2x =$ second side $2x + 1 =$ third side

Step 2: Use the equation for perimeter of a triangle as follows:
sum of the length of the sides = perimeter of a triangle.
$x + 2x + 2x + 1 = 26$

Step 3: Solve for x. $5x + 1 = 26$ so $x = 5$

Solution: first side $x = 5$ second side $2x = 10$ third side $2x + 1 = 11$

Solve the following word problems.

1. The length of a rectangle is 6 times longer than the width. The perimeter is 42. What is the width?
2. The length of a rectangle is 4 more than twice the width. The perimeter is 28. What is the length?
3. The perimeter of a triangle is 24 feet. The second side is two feet longer than the first. The third side is two feet longer then the second. What are the lengths of the sides?
4. In an isosceles triangle, two sides are equal. The third side is two less than twice the length of the two sides. The perimeter is 38. What are the lengths of the three sides?
5. The sum of the measures of the angles of a triangle is 180°. The second angle is twice the measure of the first angle. The third angle is three times the measure of the second angle. Find the measure of each angle.

5.2 Age Problems

Example 4: Tara is twice as old as Gwen. Their sister, Amy, is 5 years older than Gwen. If the sum of their ages is 29 years, find each of their ages.

Step 1: We want to find each of their ages so there are three unknowns. Tara is twice as old as Gwen, and Amy is older than Gwen, so Gwen is the youngest. Let x be Gwen's age. From the problem we can see that:

$$\left.\begin{aligned} \text{Gwen} &= x \\ \text{Tara} &= 2x \\ \text{Amy} &= x+5 \end{aligned}\right\} \text{The sum of their ages is 29.}$$

Step 2: Set up the equation, and solve for x.

$$\begin{aligned} x+2x+x+5 &= 29 \\ 4x+5 &= 29 \\ 4x &= 29-5 \\ x &= \frac{24}{4} \\ x &= 6 \end{aligned}$$

Solution:

Gwen's age $(x) = 6$
Tara's age $(2x) = 12$
Amy's age $(x+5) = 11$

Solve the following age problems.

1. Carol is 25 years older than her cousin Amanda. Cousin Bill is 3 times as old as Amanda. The sum of their ages is 90. Find each of their ages.

2. Derrick is 5 years less than twice as old as Brandon. The sum of their ages is 43. How old are Derrick and Brandon?

3. Beth's mom is 6 times older than Beth. Beth's dad is 7 years older than Beth's mom. The sum of their ages is 72. How old are each of them?

4. Delores is 2 years more than three times as old as her son, Raul. If the difference between their ages is 26, how old are Delores and Raul?

5. Eileen is 6 years older than Karen. John is three times as old as Karen. The sum of their ages is 56. How old are Eileen, Karen, and John?

6. Taylor is 18 years younger than Jim. Andrew is twice as old as Taylor. The sum of their ages is 26. How old are Taylor, Jim, and Andrew?

The following problems work in the same way as the age problems. There are two or three items of different weight, distance, number, or size. You are given the total and asked to find the amount of each item.

7. Three boxes have a total height of 640 pounds. Box A weighs twice as much as Box B. Box C weighs 30 pounds more than Box A. How much do each of the boxes weigh?

8. There are 158 students registered for American History classes. There are twice as many students registered in second period as first period. There are 10 less than three times as many students registered in third period as in first period. How many students are in each period?

9. Mei earns \$2 less than three times as much as Olivia. Shane earns twice as much as Mei. Together they earn \$594 per week. How much does each person earn per week?

10. Ellie, the elephant, eats 4 times as much as Popcorn, the pony. Zac, the zebra, eats twice as much as Popcorn. Altogether, they eat 490 kilograms of feed per week. How much feed does each of them require each week?

11. The school cafeteria served three kinds of lunches today to 225 students. The students chose the cheeseburgers three times more often than the grilled cheese sandwiches. There were twice as many grilled cheese sandwiches sold as fish sandwiches. How many of each kind of lunch were served?

12. Three friends drove west into Illinois. Kyle drove half as far as Jamaal. Conner drove 4 times as far as Kyle. Altogether, they drove 357 miles. How far did each friend drive?

13. Bianca is taking collections for this year's Feed the Hungry Project. So far she has collected \$200 more than company A than from Company B and \$800 more from Company C than from Company A. She has collected a total of \$3, 000. How much did Company C give?

14. For his birthday, Torin got \$25.00 more from his grandmother than from his uncle. His uncle gave him \$10.00 less than his cousin. Torin received \$290.00 in total. How much did he receive from his cousin?

15. Cassidy loves black and yellow jelly beans. She noticed when she was counting them that she had 8 less than three times as many black jelly beans as she had yellow jelly beans. In total, she counted 348 jelly beans. How many black jelly beans did she have?

16. Mrs. Vargus planted a garden with red and white rose bushes. Because she was studying to be a botanist, she counted the number of blossoms on each bush. She counted 5 times as many red blossoms as white blossoms. In total, she counted 1, 680 blossoms. How many red blossoms did she count?

5.3 Mixture Word Problems

When a coffee manufacturer buys coffee from two different sources at two different prices and then combines them to make a blend he or she can use algebra to design a mixture.

The formula for mixture problems is $V = AC$

V = Value of an ingredient
A = Amount of an ingredient
C = Cost per unit of the ingredient

Example 5: A coffee manufacturer bought some beans from Columbia for \$5.00 per pound and some beans from Jamaica for \$3.00 per pound. He wants to make 10 pounds of a mixture that will cost him \$4.75 per pound. How many pounds of each coffee should he use?

Step 1: Let x = The amount of \$5.00 coffee from Columbia, $10 - x$ = The amount of \$3.00 coffee from Jamaica, $x + (10 - x) = 10$ pounds of coffee blend

Step 2: Multiply each amount of coffee by its unit cost.

	Amount	×	Unit Cost	=	Value
\$5 coffee	x	×	\$5	=	$5x$
\$3 coffee	$10 - x$	×	\$3	=	$3(10 - x)$
\$4.75 coffee	10	×	\$4.75	=	$4.75(10)$

added together equals

The Value of the \$5.00 coffee + \$3.00 coffee = Value of the \$4.75 blend.
In algebra, it looks like this: $5x + 3(10 - x) = 4.75(10)$

Step 3: Solve:

$$\begin{aligned} 5x + 3(10 - x) &= 4.75(10) \\ 5x + 30 - 3x &= 47.5 \\ 2x + 30 - 30 &= 47.5 - 30 \\ \frac{2x}{2} &= \frac{17.5}{2} \\ x &= 8.75 \end{aligned}$$

Solution: Substitute the value for x, 8.75, in the equation above. The manufacturer must use 8.75 pounds of \$5.00 coffee and 1.25 pounds of \$3.00 coffee to get 10 pounds of coffee at \$4.75 per pound.

Solve the following mixture problems.

1. A meat distributor paid \$2.50 per pound for hamburger and \$4.50 per pound for ground sirloin. How many pounds of each did he use to make 100 pounds of meat mixture that will cost \$3.24 per pound?

2. How many pounds of walnuts, which cost \$4 per pound, must be mixed with 25 pounds of almonds costing \$7.50 per pound to make a mixture which will cost \$6.50 per pound?

3. In the gourmet cheese shop, employees grated cheese costing \$5.20 per pound to mix with 10 pounds of cheese which cost \$3.60 per pound to make a grated cheese topping costing \$4.80 per pound. How many pounds of cheese costing \$5.20 did they use?

4. A 200-pound bin of animal feed sells for \$1.25 per pound. How many pounds of feed costing \$2.50 per pound should be mixed with it to make a mixture which costs \$2.00 per pound?

5. A grocer mixed grape juice which costs \$2.25 per gallon with cranberry juice which costs \$1.75 per gallon. How many gallons of each should be used to make 200 gallons of cranberry/grape juice which will cost \$2.10 per gallon?

5.4 Percent Mixture Problems

Example 6: A goldsmith has 12 grams of a 40% gold alloy (alloy means it has been mixed with other metals). How many grams of pure gold should be added to make an alloy which is 65% gold?

Step 1: Make a chart for the amount of gold, A, and the percent of gold concentration, r. $Ar = Q$. Q is the quantity of the substance. Let the unknown quantity of pure gold equal x.

	Amount, A	$\times$	Concentration, r	$=$	Quantity, Q	
100% gold	x	$\times$	1.00	$=$	$1.00x$	added together
40% gold	12	$\times$	0.40	$=$	$0.40\,(12)$	
65% gold	$x+12$	$\times$	0.65	$=$	$0.65\,(x+12)$	← equals

Step 2: Solve:

$$\begin{aligned} x + 0.4\,(12) &= 0.65\,(x+12) \\ x + 4.8 &= 0.65x + 7.8 \\ 0.35x &= 3 \\ x &= 8.57 \text{ grams (round to 2 decimal places)} \end{aligned}$$

Solution: 8.57 grams of 100% pure gold must be added to make a 65% gold alloy.

Check: Replace x in the chart above to check.

100% gold: $8.57\,(1.00) = 8.57$

40% gold: $12(.40) = 4.8$

65% gold: $0.65\,(8.57 + 12) = 13.37$

equation: $8.57 + 4.8 = 13.37$

Solve the following percent mixture problems.

1. How many gallons of a 10% ammonia solution should be mixed with 50 gallons of a 30% ammonia solution to make a 15% ammonia solution?
2. How many gallons of a 25% alcohol solution must be mixed with 10 gallons of a 50% alcohol solution to make 30 gallons of a 40% alcohol solution?
3. Debbie is mixing orange juice concentrate for her restaurant. One juice concentrate is 64% real orange juice. The other is only 48% real orange juice. How many ounces of 48% real orange juice should she use to make 160 ounces of 58% real juice?
4. How many gallons of 60% antifreeze should be mixed with 40% antifreeze to make 80 gallons of 45% antifreeze?
5. Hank has some 60% maple syrup and some 100% maple syrup in his restaurant. How many ounces of each should he use to make 100 ounces of 85% maple syrup?
6. A butcher has some hamburger which is 4% fat and some hamburger which is 20% fat. How much of each will he need to make 120 pounds of hamburger which is 10% fat?

5.5 Working Together Problems

Example 7: If Barbara can do a certain job in 4 hours, and Kelly can do the same job in 6 hours, how long would it take them to do the job if they worked together?

Caution: At first glance, you may want to just average the two times together and conclude that Barbara and Kelly could do the job together in 5 hours. But, think about this problem carefully. If Barbara can do the job by herself in 4 hours, the job would be done even faster than 4 hours if she had Kelly's help, even if Kelly doesn't work as fast as Barbara. So, the answer you are looking for should be less than 4 hours.

Step 1: Let x = the amount of time it takes them to do the job together. Barbara can do $\frac{1}{4}$ of the job in 1 hour, and Kelly can do $\frac{1}{6}$ of the job in 1 hour. Make a chart of what you know:

	Barbara +	Kelly =	Together
Number of tasks	1 job	1 job	1 job
Time	4 hours	6 hours	x hours

Step 2: Write an equation and solve: $\frac{1 \text{ job}}{4 \text{ hours}} + \frac{1 \text{ job}}{6 \text{ hours}} = \frac{1 \text{ job}}{x \text{ hours}}$

Find a common denominator: $\frac{3}{12} + \frac{2}{12} = \frac{1}{x} \longrightarrow \frac{5}{12} = \frac{1}{x}$

Solve by cross multiplying: $5x = 12$, so $x = \frac{12}{5}$ or $2\frac{2}{5}$

Solution: Kelly and Barbara can do the job together in $2\frac{2}{5}$ hours.

Example 8: Jerome and Kristen are conducting a hand recount of city election ballots for the office of mayor. Jerome can count 240 ballots per hour while Kristen can count 320 ballots per hour. How long will it take for Kristen and Jerome to count 840 votes if they work together?

Let x = the number of hours the job takes if they work together.

	Jerome +	Kristen =	Together
Number of tasks	240 ballots	320 ballots	840 ballots
Time	1 hour	1 hour	x hours

$$\frac{240 \text{ ballots}}{1 \text{ hour}} + \frac{320 \text{ ballots}}{1 \text{ hour}} = \frac{840 \text{ ballots}}{x \text{ hours}} \longrightarrow \frac{560}{1} = \frac{840}{x}$$

$$560x = 840, \qquad x = \frac{840}{560} \text{ or } 1\frac{1}{2}$$

Solution: Jerome and Kristen can count 840 ballots in $1\frac{1}{2}$ hours.

Solve the following working together problems.

1. Simone can assemble a radio in 4 hours, and Sheila can assemble the same model radio in 7 hours. How long would it take them to assemble one radio if they worked together?
2. John can change the front brake pads on a car in 3 hours. Manuel can change the front brake pads in 2 hours. How long would it take John and Manuel to change the brake pads on a car if they worked together?
3. Jessica can type 4 pages per hour. Her friend, Alejandro, can type 6 pages per hour. How long would it take to type a 14-page paper if they worked together?
4. Sandra and Michael both work at a restaurant as servers. Sandra is able to serve six tables of people per hour, while Michael is able to serve 4 tables per hour. If both of the servers work at a private party at the restaurant serving 13 tables of people, how long will it take them to serve the tables?
5. Diedra can paint a wall mural every three months. Lucy can paint a wall mural every five months. How long would it take them to paint one wall mural if they worked together?
6. Phillip can bake 12 dozen cookies per hour. Brian can bake one dozen cookies in the same amount of time. How long would it take them to bake 30 dozen cookies if they both worked together?

5.6 Consecutive Integer Problems

	Examples:	Algebraic notation:
Consecutive integers follow each other in order	$1, 2, 3, 4$ $-3, -4, -5, -6$	$n, n+1, n+2, n+3$
Consecutive **even** integers:	$2, 4, 6, 8, 10$ $-12, -14, -16, -18$	$n, n+2, n+4, n+6$
Consecutive **odd** integers:	$3, 5, 7, 9$ $-5, -7, -9, -11$	$n, n+2, n+4, n+6$

Example 9: The sum of three consecutive odd integers is 63. Find the integers.

Step 1: Represent the three odd integers:
Let n = the first odd integer
$n+2$ = the second odd integer
$n+4$ = the third odd integer

Step 2: The sum of the integers is 63, so the algebraic equation is
$n + n + 2 + n + 4 = 63$. Solve for n.
$n = 19$

Solution: the first odd integer = 19
the second odd integer = 21
the third odd integer = 23

Check: Does $19 + 21 + 23 = 63$? Yes, it does.

Example 10: Find three consecutive odd integers such that the sum of the first and second is three less than the third.

Step 1: Represent the three odd integers just like example 9:
Let n = the first odd integer
$n + 2$ = the second odd integer
$n + 4$ = the third odd integer

Step 2: In this problem, the sum of the first and second integers is three less than the third integer, so the algebraic equation is written as follows:
$n + n + 2 = n + 4 - 3$
$n = -1$

Solution: the first odd integer $= -1$
the second odd integer $= 1$
the third odd integer $= 3$

Check: Is the sum of -1 and 1 three less than 3?
$-1 + 1 = 3 - 3$ or $0 = 0$ Yes, it is.

Solve the following problems.

1. Find three consecutive odd integers whose sum is 141.

2. Find three consecutive integers whose sum is -21.

3. The sum of three consecutive even integers is 48. What are the numbers?

4. Find two consecutive even integers such that six times the first equals five times the second.

5. Find two consecutive odd integers such that seven times the first equals five times the second.

6. Find two consecutive odd numbers whose sum is forty-four.

5.7 Inequality Word Problems

Inequality word problems involve staying under a limit or having a minimum goal one must meet.

Example 11: A contestant on a popular game show must earn a minimum of 800 points by answering a series of questions worth 40 points each per category in order to win the game. The contestant will answer questions from each of four categories. Her results for the first three categories are as follows: 160 points, 200 points, and 240 points. Write an inequality which describes how many points, (p), the contestant will need on the last category in order to win.

Step 1: Add to find out how many points she already has. $160 + 200 + 240 = 600$

Step 2: Subtract the points she already has from the minimum points she needs. $800 - 600 = 200$. She must earn at least 200 points in the last category to win. If she earns more than 200 points, that is okay, too. To express the number of points she needs, use the following inequality statement:

$p \geq 200$ The points she needs must be greater than or equal to 200.

Solve each of the following problems using inequalities.

1. Stella wants to place her money in a high-interest money market account. However, she needs at least $1000 to open an account. Each month, she sets aside some of her earnings in a savings account. In January through June, she added the following amounts to her savings: $121, $206, $138, $212, $109, and $134. Write an inequality which describes the amount of money she can set aside in July to qualify for the money market account.

2. A high school band program will receive $2,000.00 for selling $10,000.00 worth of coupon books. Six band classes participate in the sales drive. Classes 1–5 collect the following amounts of money: $1,400, $2,600, $1,800, $2,450, and $1,550. Write an inequality which describes the amount of money the sixth class must collect so that the band will receive $2,000.

3. A small elevator has a maximum capacity of 1,000 pounds before the cable holding it in place snaps. Six people get on the elevator. Five of their weights follow: 146, 180, 130, 262, and 135. Write an inequality which describes the amount the sixth person can weigh without snapping the cable.

4. A small high school class of 9 students were told they would receive a pizza party if their class average was 92% or higher on the next exam. Students 1–8 scored the following on the exam: 86, 91, 98, 83, 97, 89, 99, and 96. Write an inequality which describes the score the ninth student must make for the class to qualify for the pizza party.

5. Raymond wants to spend his entire credit limit on his credit card. His credit limit is $2,000. He purchases items costing $600, $800, $50, $168, and $3. Write an inequality which describes the amounts Raymond can put on his credit card for his next purchases.

Chapter 5 Review

1. Deanna is five years more than six times older than Ted. The sum of their ages is 47. How old is Ted?

2. Ross is six years older than twice his sister Holly's age. The difference is their ages is 18 years. How old is Holly?

3. The band members sold tickets to their concert performance. Some were \$5 tickets, and some were \$6 tickets. There were 16 more than twice as many \$6 tickets sold as \$5 tickets. The total sales were \$1643. How many tickets of each price were sold?

4. Three consecutive integers have a sum of 240. Find the integers.

5. Find three consecutive even numbers whose sum is negative seventy-two.

6. The sum of two numbers is 55. The larger number is 7 more than twice the smaller number. What are the numbers?

7. One number is 10 more than the other number. Twice the smaller number is 13 more than the larger number. What are the numbers?

8. The perimeter of a triangle is 43 inches. The second side is three inches longer than the first side. The third side is one inch longer than the second. Find the length of each side.

9. The perimeter of a rectangle is 80 feet. The length of the rectangle is 2 feet less than 5 times the width. What is the length and width of the rectangle?

10. Priscilla bought 20 pounds of chocolate-covered peanuts for \$3.50 per pound. How many pounds of chocolate-covered walnuts would she have to buy at \$7.00 per pound to make boxes of nut mixtures that would cost her \$4.50 per pound?

11. One solution is 15% ammonia. A second solution is 40% ammonia. How many ounces of each should be used to make 100 ounces of a 20% ammonia solution?

12. Joe, Craig, and Dylan have a combined weight of 429 pounds. Craig weighs 34 pounds more than Joe. Dylan weighs 13 pounds more than Craig. How many pounds does Craig weigh?

13. Jesse and Larry entered a pie-eating contest. Jesse ate 2 less than twice as many pies as Larry. They ate a total of 28 pies. How many pies did Larry eat?

14. Lena and Jodie are sisters, and together they have 68 bottles of nail polish. Lena bought 5 more than half the bottles. How many did Jodie buy?

15. Janet and Artie wanted to play tug-of-war. Artie pulls with 150 pounds of force while Janet pulls with 40 pounds of force. In order to make this a fair contest, Janet enlists the help of her friends Trudi, Sherri, and Bridget who pull with 30, 25, and 40 pounds respectively. Write an inequality describing the minimum amount Janet's fourth friend, Tommy, must pull to beat Artie.

16. Jim takes great pride in decorating his float for the homecoming parade for his high school. With the \$5,000 he has to spend, Jim bought 5,000 carnations at \$.25 each, 4,000 tulips at \$.50 each, and 300 irises at \$.90 each. Write an inequality which describes how many roses, r, Jim can buy if roses cost \$.80 each.

Chapter 6
Matrices

This chapter covers the following North Carolina mathematics standards for Algebra I:

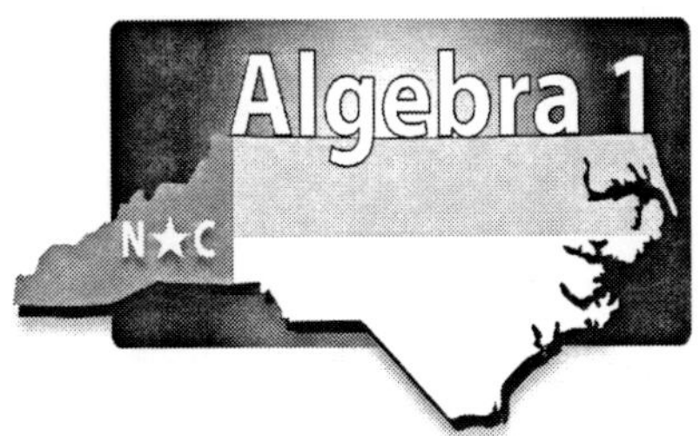

Competency Goal	Objectives
Number and Operations	1.01 1.02
Data Analysis and Probability	3.01 3.02 3.03b
Algebra	4.01

A **matrix** (plural: **matrices**) is an ordered array of numbers, and each number in a matrix is called an **element**. The matrix shown below contains the elements 3, -1, 2, 4, 0, and 1. It is arranged in two rows and three columns and, therefore, is referred to as a 2×3 matrix. When describing a matrix, always give the number of rows and then the number of columns.

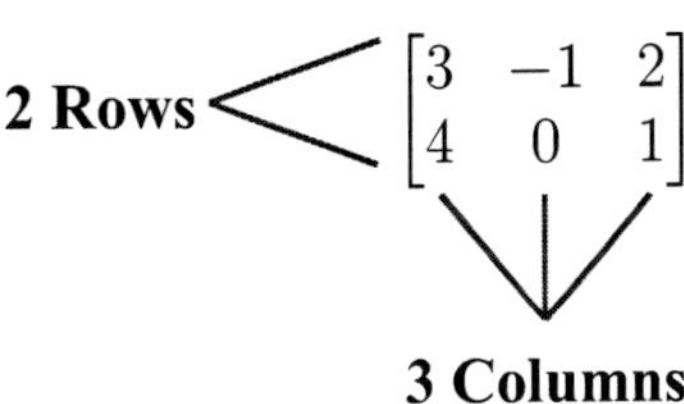

$$\begin{bmatrix} 3 & -1 & 2 \\ 4 & 0 & 1 \end{bmatrix}$$

Matrices can be used to solve systems of linear equations, which are covered later in this book. Since their invention in 1858, matrices have played a role in such fields as economics, engineering, and quantum mechanics. This chapter will cover the use of basic operations on matrices, but will not address solving systems of linear equations using matrices. To learn how to solve systems of linear equations using matrices, consult an algebra and trigonometry textbook.

6.1 Addition of Matrices

Matrices to be added must be of the same size; they need to have the same number of rows and columns. A 2×2 matrix can only be added to another 2×2 matrix. A 2×3 matrix can only be added to another 2×3 matrix, but **cannot** be added to a 3×2 matrix.

To add two matrices of the same size, add the corresponding elements of the two matrices. The resulting matrix is the same size as each of the two matrices that were added together.

Example 1: $\begin{bmatrix} 7 & -2 \\ -1 & 4 \end{bmatrix} + \begin{bmatrix} -6 & 2 \\ 4 & 0 \end{bmatrix} = \begin{bmatrix} 7+(-6) & (-2)+2 \\ (-1)+4 & 4+0 \end{bmatrix} = \begin{bmatrix} 1 & 0 \\ 3 & 4 \end{bmatrix}$

Note that the resulting matrix is a 2×2 matrix, as are the matrices added together.

Example 2: $\begin{bmatrix} 7 & -2 \\ -1 & 4 \\ 5 & -3 \end{bmatrix} + \begin{bmatrix} -6 & 2 & 2 \\ 4 & 0 & 0 \end{bmatrix} =$

The matrices to be added are not of the same size. The first matrix is a 3×2 matrix, and the second matrix is 2×3. Therefore, these two matrices cannot be added. It is not possible to add these matrices.

Note: Any number of matrices of the same size can be added together.

Add the matrices together when possible. When the matrices cannot be added, write NP.

1. $\begin{bmatrix} 8 & 4 \\ 5 & -3 \end{bmatrix} + \begin{bmatrix} 0 & -7 \\ -4 & -9 \end{bmatrix}$

2. $\begin{bmatrix} 6 \\ -4 \\ 5 \end{bmatrix} + \begin{bmatrix} 2 \\ -9 \\ -8 \end{bmatrix} + \begin{bmatrix} 9 \\ -1 \\ -3 \end{bmatrix} + \begin{bmatrix} 5 \\ -1 \\ -2 \end{bmatrix}$

3. $\begin{bmatrix} 3 & -1 & 2 \\ 4 & 0 & 1 \end{bmatrix} + \begin{bmatrix} 6 & 0 & 5 \\ 4 & 0 & 1 \end{bmatrix}$

4. $\begin{bmatrix} -6 \\ -2 \end{bmatrix} + \begin{bmatrix} -4 & -1 \end{bmatrix}$

5. $\begin{bmatrix} -5 & -2 \\ 9 & -7 \\ 3 & 6 \end{bmatrix} + \begin{bmatrix} -1 & -2 \\ -8 & 7 \\ 9 & -4 \end{bmatrix}$

6. $\begin{bmatrix} 8 & -1 & -6 & 3 \\ 0 & -7 & -5 & -4 \end{bmatrix} + \begin{bmatrix} -8 & 1 & 5 & -2 \\ 1 & 6 & 6 & 5 \end{bmatrix}$

6.2 Multiplication of a Matrix by a Constant

A matrix can be multiplied by a constant. The constant is multiplied by each element in the matrix. The resulting matrix is the same size as the matrix being multiplied.

Example 3: $4\begin{bmatrix}1 & 0\\ 3 & 4\end{bmatrix} =$

Step 1: Multiply every number in the matrix by 4.

$$\begin{bmatrix}(4\times 1) & (4\times 0)\\ (4\times 3) & (4\times 4)\end{bmatrix} = \begin{bmatrix}4 & 0\\ 12 & 16\end{bmatrix}$$

Step 2: The solution of $4\begin{bmatrix}1 & 0\\ 3 & 4\end{bmatrix}$ is $\begin{bmatrix}4 & 0\\ 12 & 16\end{bmatrix}$.

Example 4: $-3\begin{bmatrix}-8 & 1 & 5 & -2\\ 1 & 6 & 6 & 5\end{bmatrix} =$

Step 1: Multiply every number in the matrix by -3.

$$\begin{bmatrix}(-3\times -8) & (-3\times 1) & (-3\times 5) & (-3\times -2)\\ (-3\times 1) & (-3\times 6) & (-3\times 6) & (-3\times 5)\end{bmatrix} = \begin{bmatrix}24 & -3 & -15 & 6\\ -3 & -18 & -18 & -15\end{bmatrix}$$

Step 2: The solution of $-3\begin{bmatrix}-8 & 1 & 5 & -2\\ 1 & 6 & 6 & 5\end{bmatrix}$ is $\begin{bmatrix}24 & -3 & -15 & 6\\ -3 & -18 & -18 & -15\end{bmatrix}$

Multiply each of the following.

1. $-2\begin{bmatrix}-8 & 1\\ 6 & 7\\ -5 & -4\end{bmatrix}$

2. $5\begin{bmatrix}3\\ 0\\ -1\end{bmatrix}$

3. $\frac{1}{2}\begin{bmatrix}2 & -4 & -9 & 7\\ -1 & 6 & 3 & -8\end{bmatrix}$

4. $-\frac{1}{4}\begin{bmatrix}-1 & -6 & 4 & -2\\ 8 & 0 & 3 & -8\\ 4 & 6 & -3 & 12\end{bmatrix}$

5. $-9\begin{bmatrix}-10 & -1 & \frac{1}{3} & 7 & -\frac{3}{4}\end{bmatrix}$

6. $6\begin{bmatrix}-4 & -5\\ 1 & 0\end{bmatrix}$

6.3 Subtraction of Matrices

Subtraction of matrices is similar to addition of matrices in that the matrices to be subtracted must be the same size. Suppose x and y represent two different matrices of the same size. $x - y$ can also be written $x + (-1)\,y$. Therefore, subtraction of matrices involves two steps: multiplying the second matrix by -1 and then adding it to the first matrix.

Example 5: $\begin{bmatrix} 4 & -6 \\ 9 & -5 \end{bmatrix} - \begin{bmatrix} 3 & -1 \\ -4 & 7 \end{bmatrix}$ can also be written as $\begin{bmatrix} 4 & -6 \\ 9 & -5 \end{bmatrix} + (-1)\begin{bmatrix} 3 & -1 \\ -4 & 7 \end{bmatrix}$

Step 1: Multiply the second matrix by -1.

$$(-1)\begin{bmatrix} 3 & -1 \\ -4 & 7 \end{bmatrix} = \begin{bmatrix} -3 & 1 \\ 4 & -7 \end{bmatrix}$$

Step 2: Add the first matrix and the product from step 1.

$$\begin{bmatrix} 4 & -6 \\ 9 & -5 \end{bmatrix} + \begin{bmatrix} -3 & 1 \\ 4 & -7 \end{bmatrix} = \begin{bmatrix} 1 & -5 \\ 13 & -12 \end{bmatrix}$$

Subtract the matrices when possible. When the matrices cannot be subtracted, write NP.

1. $\begin{bmatrix} 1 & 2 \\ 6 & 0 \\ -1 & 4 \end{bmatrix} - \begin{bmatrix} 3 & 2 \\ 1 & -3 \\ 5 & 1 \end{bmatrix}$

2. $\begin{bmatrix} 1 & 2 & 3 \\ 4 & 5 & 6 \end{bmatrix} - \begin{bmatrix} 3 & 2 & 1 \\ 5 & 4 & 6 \end{bmatrix}$

3. $\begin{bmatrix} 6 & 2 & 2 \\ -4 & -3 & -9 \\ 5 & 1 & -8 \end{bmatrix} - \begin{bmatrix} 6 & 9 & -4 \\ -4 & 0 & -8 \\ 2 & 3 & -7 \end{bmatrix}$

4. $\begin{bmatrix} 2 & 3 \\ 1 & 5 \end{bmatrix} - \begin{bmatrix} 1 & 3 \\ -2 & 1 \end{bmatrix}$

5. $\begin{bmatrix} 8 & -1 & -6 & 3 \\ 0 & -7 & -5 & -4 \end{bmatrix} - \begin{bmatrix} -8 & 1 & 5 & -2 \\ 1 & 6 & 6 & 5 \end{bmatrix}$

6. $\begin{bmatrix} 1 & 2 & 5 & 5 \\ 7 & 9 & 1 & -8 \\ -3 & 2 & 5 & -7 \end{bmatrix} - \begin{bmatrix} 1 & 3 & 1 \\ -1 & 0 & -8 \\ 0 & 1 & -7 \end{bmatrix}$

Perform the proper operation(s) on each set of matrices.

7. $\begin{bmatrix} 4 & -1 & 2 & 7 \\ -3 & 1 & -5 & -4 \end{bmatrix} + \frac{1}{2}\begin{bmatrix} 6 & 8 & 10 & -2 \\ -4 & 2 & 0 & -6 \end{bmatrix}$

8. $4\begin{bmatrix} -1 & 2 & -\frac{1}{2} & 0 & -\frac{3}{4} \end{bmatrix} - \begin{bmatrix} -1 & 0 & -5 & 6 & -3 \end{bmatrix}$

6.4 Applications with Matrices

Example 6: Find a, b, c, and d such that

$$\begin{bmatrix} 2 & -1 \\ 4 & 3 \end{bmatrix} + \begin{bmatrix} a & b \\ c & d \end{bmatrix} = \begin{bmatrix} 3 & -1 \\ 2 & 2 \end{bmatrix}$$

Step 1: Write the equation for the four sets of corresponding elements.

$2 + a = 3 \qquad -1 + b = -1$

$4 + c = 2 \qquad 3 + d = 2$

Step 2: Solve each of the four equations.

$a = 1 \qquad b = 0$

$c = -2 \qquad d = -1$

Example 7: Lucie and her friend Laura are shopping for a new cellular phone plan. Plan A offers 400 minutes per month for \$60, plus another 200 night and weekend minutes for an extra \$20. Plan B offers 500 monthly minutes for \$50, plus 150 night and weekend minutes for an extra \$20. The two plans can be represented in the following matrices:

$$\begin{bmatrix} \$60 & \$20 \\ 400 & 200 \end{bmatrix} = \text{A} \qquad \begin{bmatrix} \$50 & \$20 \\ 500 & 150 \end{bmatrix} = \text{B}$$

If Lucie chooses Plan A and Laura chooses Plan B, what will the total cost of their services be if they also select the night and weekend minutes, and how many total minutes will they receive? What is the average cost and minutes of the plans?

Step 1: To determine the total of the two plans, add the two matrices together, A + B.

$$\begin{bmatrix} \$60 + \$50 & \$20 + \$20 \\ 400 + 500 & 500 + 150 \end{bmatrix} = \begin{bmatrix} \$110 & \$40 \\ 900 & 350 \end{bmatrix}$$

Step 2: Calculate the average cost and minutes by multiplying the total matrix by $\frac{1}{2}$ (or dividing it by 2):

$$\frac{1}{2} \times \begin{bmatrix} \$110 & \$40 \\ 900 & 350 \end{bmatrix} = \begin{bmatrix} \$55 & \$20 \\ 450 & 175 \end{bmatrix}$$

Solve the following matrix problems.

1. $\begin{bmatrix} d & 3 \\ e & 1 \end{bmatrix} + \begin{bmatrix} 2 & f \\ 2 & g \end{bmatrix} = \begin{bmatrix} 5 & 3 \\ 1 & 2 \end{bmatrix}$

2. $\begin{bmatrix} 3d & 1 \\ 2 & 3g \\ 1 & f \end{bmatrix} - \begin{bmatrix} 2d & 0 \\ e & g \\ 1 & 2 \end{bmatrix} = \begin{bmatrix} -2 & 1 \\ 4 & 4 \\ 0 & 1 \end{bmatrix}$

3. $3\begin{bmatrix} 1 & 0 \\ 2 & 1 \end{bmatrix} + \frac{1}{2}\begin{bmatrix} 4 & 6 \\ 0 & 2 \end{bmatrix} = \begin{bmatrix} d & f \\ e & g \end{bmatrix}$

4. A computer company with one plant in the West and one plant in the East produces monitors and printers. The production for January and February are give as follows:

	West Plant	East Plant		West Plant	East Plant	
January =	2000	1710	February =	2300	1850	Monitors
	800	650		950	800	Printers

(A) What is the average monthly production of the monitors and printers?
(B) What is the increase from January to February?
(C) What is the total production for January and February?

5. The Yummy Candy Company produces a variety of candy products and packages them for various holidays. The Christmas package consists of three pieces of chocolate, two pecan candies, one peppermint twist, and four chocolate-covered cherries. The Valentine package consists of the same package, but contains three times as many pieces of each candy. Write the number of candies in both the Christmas and Valentine packages in matrix form.

6. What are the total numbers of candies contained in one Christmas package and one Valentine package from problem 5?

Chapter 6 Review

Solve each matrix. If not possible, write not possible.

1. $2\begin{bmatrix} -5 & -1 \\ 4 & -7 \\ 9 & 6 \end{bmatrix} + \begin{bmatrix} -1 & -5 \\ -8 & 7 \\ 1 & -4 \end{bmatrix} =$

2. $\begin{bmatrix} 4 \\ -1 \end{bmatrix} + \begin{bmatrix} -7 & -1 \end{bmatrix} =$

3. $\begin{bmatrix} 9 & -1 & -6 & 3 \\ 0 & -5 & -5 & -4 \end{bmatrix} + \begin{bmatrix} -8 & 1 & 2 & -3 \\ 1 & 6 & 11 & 5 \end{bmatrix} =$

4. $-\frac{1}{4}\begin{bmatrix} -1 & -6 & 4 & -8 \\ 12 & 0 & 3 & -2 \\ 4 & 6 & -4 & 16 \end{bmatrix} =$

Subtract the following matrices. If not possible, write not possible.

5. $\begin{bmatrix} 2 & 4 \\ 0 & 5 \end{bmatrix} - \begin{bmatrix} 1 & 7 \\ -3 & 1 \end{bmatrix} =$

6. $\begin{bmatrix} 10 & -1 & -6 & 3 \\ 0 & -5 & -5 & -4 \end{bmatrix} - \begin{bmatrix} -8 & 9 & 1 & -2 \\ 1 & 0 & 6 & 4 \end{bmatrix} =$

7. $\begin{bmatrix} 1 & 2 & 5 & 4 \\ 7 & 0 & 1 & -8 \\ -3 & 2 & 5 & -2 \end{bmatrix} - \begin{bmatrix} 1 & 3 & 13 \\ -1 & 0 & -11 \\ 0 & 5 & -7 \end{bmatrix} =$

8. Mr. Thompson goes on two road trips per year. His two favorite places to go are Dallas, TX and Atlantic City, NJ. When he went to Dallas last year, he spent $3,000 and drove 1,235 miles, and when he went to Atlantic City last year, he spent $5,500 and drove 786 miles. This year when he goes on vacation, he will have $4,300 to spend in Dallas and $4,900 in Atlantic City. He will drive 200 more miles because he is picking up his sister for both trips.
   (A) Write two 2×2 matrices that include miles and price for the Dallas trip and Atlantic City trip.
   (B) What is the difference in the amount of money Mr. Thompson will spend this year compared to last year?

For questions 9 and 10, find a, b, c, and d.

9. $\begin{bmatrix} a & 4 \\ b & 5 \end{bmatrix} + \begin{bmatrix} 2 & c \\ 0 & d \end{bmatrix} = \begin{bmatrix} 5 & 3 \\ 1 & 7 \end{bmatrix}$

10. $5\begin{bmatrix} 8 & 0 \\ 1 & 3 \end{bmatrix} + \frac{1}{2}\begin{bmatrix} 10 & 8 \\ 0 & 2 \end{bmatrix} = \begin{bmatrix} a & c \\ b & d \end{bmatrix}$

Chapter 7
Polynomials

This chapter covers the following North Carolina mathematics standards for Algebra I:

Competency Goal	Objectives
Number and Operations	1.01b

Polynomials are algebraic expressions which include **monomials** containing one term, **binomials** which contain two terms, and **trinomials**, which contain three terms. Expressions with more than three terms are called **polynomials. Terms** are separated by plus and minus signs.

Examples

Monomials	**Binomials**	**Trinomials**	**Polynomials**
$4f$	$4t + 9$	$x^2 + 2x + 3$	$x^3 - 3x^2 + 3x - 9$
$3x^3$	$9 - 7g$	$5x^2 - 6x - 1$	$p^4 + 2p^3 + p^2 - 5 + p9$
$4g^2$	$5x^2 + 7x$	$y^4 + 15y^2 + 100$	
2	$6x^3 - 8x$		

7.1 Adding and Subtracting Monomials

Two **monomials** are added or subtracted as long as the **variable and its exponent** are the **same**. This is called combining like terms. Use the same rules you used for adding and subtracting integers.

Example 1: $4x + 5x = 9x$ $\quad \begin{array}{r} 3x^4 \\ -8x^4 \\ \hline -5x^4 \end{array}$ $\quad 2x^2 - 9x^2 = -7x^2$ $\quad \begin{array}{r} 5y \\ +2y \\ \hline 7y \end{array}$ $\quad 6y^3 - 5y^3 = y^3$

Remember: When the integer in front of the variable is "1", it is usually not written. $1x^2$ is the same as x^2, and $-1x$ is the same as $-x$.

Add or subtract the following monomials.

1. $2x^2 + 5x^2$
2. $5t + 8t$
3. $9y^3 - 2y^3$
4. $6g - 8g$
5. $7y^2 + 8y^2$
6. $s^5 + s^5$
7. $-2x - 4x$
8. $4w^2 - w^2$
9. $z^4 + 9z^4$
10. $-k + 2k$
11. $3x^2 - 5x^2$
12. $9t + 2t$
13. $-7v^3 + 10v^3$
14. $-2x^3 + x^3$
15. $10y^4 - 5y^4$

16. $\begin{array}{r} y^4 \\ +2y^4 \\ \hline \end{array}$ 18. $\begin{array}{r} 8t^2 \\ +7t^2 \\ \hline \end{array}$ 20. $\begin{array}{r} 5w^2 \\ +8w^2 \\ \hline \end{array}$ 22. $\begin{array}{r} -5z \\ +9z \\ \hline \end{array}$ 24. $\begin{array}{r} 7t^3 \\ -6t^3 \\ \hline \end{array}$

17. $\begin{array}{r} 4x^3 \\ -9x^3 \\ \hline \end{array}$ 19. $\begin{array}{r} -2y \\ -4y \\ \hline \end{array}$ 21. $\begin{array}{r} 11t^3 \\ -4t^3 \\ \hline \end{array}$ 23. $\begin{array}{r} 4w^5 \\ +w^5 \\ \hline \end{array}$ 25. $\begin{array}{r} 3x \\ +8x \\ \hline \end{array}$

7.2 Adding Polynomials

When adding **polynomials,** make sure the exponents and variables are the same on the terms you are combining. The easiest way is to put the terms in columns with **like exponents** under each other. Each column is added as a separate problem. Fill in the blank spots with zeros if it helps you keep the columns straight. You never carry to the next column when adding polynomials.

Example 2: Add $3x^2 + 14$ and $5x^2 + 2x$

$$\begin{array}{r} 3x^2 + 0x + 14 \\ (+)\, 5x^2 + 2x + 0 \\ \hline 8x^2 + 2x + 14 \end{array}$$

Example 3: $(4x^3 - 2x) + (-x^3 - 4)$

$$\begin{array}{r} 4x^3 - 2x + 0 \\ (+) - x^3 + 0x - 4 \\ \hline 3x^3 - 2x - 4 \end{array}$$

Add the following polynomials.

1. $y^2 + 3y + 2$ and $2y^2 + 4$
2. $(5y^2 + 4y - 6) + (2y^2 - 5y + 8)$
3. $5x^3 - 2x^2 + 4x - 1$ and $3x^2 - x + 2$
4. $-p + 4$ and $5p^2 - 2p + 2$
5. $(w - 2) + (w^2 + 2)$
6. $4t^2 - 5t - 7$ and $8t + 2$
7. $t^4 + t + 8$ and $2t^3 + 4t - 4$
8. $(3s^3 + s^2 - 2) + (-2s^3 + 4)$
9. $(-v^2 + 7v - 8) + (4v^3 - 6v + 4)$
10. $6m^2 - 2m + 10$ and $m^2 - m - 8$
11. $-x + 4$ and $3x^2 + x - 2$
12. $(8t^2 + 3t) + (-7t^2 - t + 4)$
13. $(3p^4 + 2p^2 - 1) + (-5p^2 - p + 8)$
14. $12s^3 + 9s^2 + 2s$ and $s^3 + s^2 + s$
15. $(-9b^2 + 7b + 2) + (-b^2 + 6b + 9)$
16. $15c^2 - 11c + 5$ and $-7c^2 + 3c - 9$
17. $5c^3 + 2c^2 + 3$ and $2c^3 + 4c^2 + 1$
18. $-14x^3 + 3x^2 + 15$ and $7x^3 - 12$
19. $(-x^2 + 2x - 4) + (3x^2 - 3)$
20. $(y^2 - 11y + 10) + (-13y^2 + 5y - 4)$
21. $3d^5 - 4d^3 + 7$ and $2d^4 - 2d^3 - 2$
22. $(6t^5 - t^3 + 17) + (4t^5 + 7t^3)$
23. $4p^2 - 8p + 9$ and $-p^2 - 3p - 5$
24. $20b^3 + 15b$ and $-4b^2 - 5b + 14$
25. $(-2w + 11) + (w^3 + w - 4)$
26. $(25z^2 + 13z + 8) + (z^2 - 2z - 10)$

7.3 Subtracting Polynomials

When you subtract polynomials, it is important to remember to change all the signs in the subtracted polynomial (the subtrahend) and then add.

Example 4: $(4y^2 + 8y + 9) - (2y^2 + 6y - 4)$

Step 1: Copy the subtraction problem into vertical form. Make sure you line up the terms with like exponents under each other.

$$\begin{array}{r} 4y^2 + 8y + 9 \\ (-)\,2y^2 + 6y - 4 \\ \hline \end{array}$$

Step 2: Change the subtraction sign to addition and all the signs of the subtracted polynomial to the opposite sign.

$$\begin{array}{r} 4y^2 + 8y + 9 \\ (+) - 2y^2 - 6y + 4 \\ \hline 2y^2 + 2y + 13 \end{array}$$

Subtract the following polynomials.

1. $(2x^2 + 5x + 2) - (x^2 + 3x + 1)$
2. $(8y - 4) - (4y + 3)$
3. $(11t^3 - 4t^2 + 3) - (-t^3 + 4t^2 - 5)$
4. $(-3w^2 + 9w - 5) - (-5w^2 - 5)$
5. $(6a^5 - a^3 + a) - (7a^5 + a^2 - 3a)$
6. $(14c^4 + 20c^2 + 10) - (7c^4 + 5c^2 + 12)$
7. $(5x^2 - 9x) - (-7x^2 + 4x + 8)$
8. $(12y^3 - 8y^2 - 10) - (3y^3 + y + 9)$
9. $(-3h^2 - 7h + 7) - (5h^2 + 4h + 10)$
10. $(10k^3 - 8) - (-4k^3 + k^2 + 5)$
11. $(x^2 - 5x + 9) - (6x^2 - 5x + 7)$
12. $(12p^2 + 4p) - (9p - 2)$
13. $(-2m - 8) - (6m + 2)$
14. $(13y^3 + 2y^2 - 8y) - (2y^3 + 4y^2 - 7y)$
15. $(7g + 3) - (g^2 + 4g - 5)$
16. $(-8w^3 + 4w) - (-10w^3 - 4w^2 - w)$
17. $(12x^3 + x^2 - 10) - (3x^3 + 2x^2 + 1)$
18. $(2a^2 + 2a + 2) - (-a^2 + 3a + 3)$
19. $(c + 19) - (3c^2 - 7c + 2)$
20. $(-6v^2 + 12v) - (3v^2 + 2v + 6)$
21. $(4b^3 + 3b^2 + 5) - (7b^3 - 8)$
22. $(15x^3 + 5x^2 - 4) - (4x^3 - 4x^2)$
23. $(8y^2 - 2) - (11y^2 - 2y - 3)$
24. $(-z^2 - 5z - 8) - (3z^2 - 5z + 5)$

7.4 Multiplying Monomials

When two monomials have the **same variable**, you can multiply them. Then, add the **exponents** together. If the variable has no exponent, it is understood that the exponent is 1.

Example 5: $4x^4 \times 3x^2 = 12x^6$ $2y \times 5y^2 = 10y^3$

Multiply the following monomials.

1. $6a \times 9a^5$
2. $2x^6 \times 5x^3$
3. $4y^3 \times 3y^2$
4. $10t^2 \times 2t^2$
5. $2p^5 \times 4p^2$
6. $9b^2 \times 8b$
7. $3c^3 \times 3c^3$
8. $2d^8 \times 9d^2$
9. $6k^3 \times 5k^2$
10. $7m^5 \times m$
11. $11z \times 2z^7$
12. $3w^4 \times 6w^5$
13. $4x^4 \times 5x^3$
14. $5n^2 \times 3n^3$
15. $8w^7 \times w$
16. $10s^6 \times 5s^3$
17. $4d^5 \times 4d^5$
18. $5y^2 \times 8y^6$
19. $7t^{10} \times 3t^5$
20. $6p^8 \times 2p^3$
21. $x^3 \times 2x^3$

When problems include negative signs, follow the rules for multiplying integers.

22. $-7s^4 \times 5s^3$
23. $-6a \times -9a^5$
24. $4x \times -x$
25. $-3y^2 \times -y^3$
26. $-5b^2 \times 3b^5$
27. $9c^4 \times -2c$
28. $-4t^3 \times 8t^3$
29. $10d \times -8d^7$
30. $-3g^6 \times -2g^3$
31. $-7s^4 \times 7s^3$
32. $-d^3 \times -2d$
33. $11p \times -2p^5$
34. $-5x^7 \times -3x^3$
35. $8z^4 \times 7z^4$
36. $-4w \times -5w^8$
37. $-5y^4 \times 6y^2$
38. $9x^3 \times -7x^5$
39. $-a^4 \times -a$
40. $-7k^2 \times 3k$
41. $-15t^2 \times -t^4$
42. $3x^8 \times 9x^2$

7.5 Multiplying Monomials by Polynomials

In the chapter on solving multi-step equations, you learned to remove parentheses by multiplying the number outside the parentheses by each term inside the parentheses: $2(4x - 7) = 8x - 14$. Multiplying monomials by polynomials works the same way.

Example 6: $-5t(2t^2 - 7t + 9)$

Step 1: Multiply $-5t \times 2t^2 = -10t^3$

Step 2: Multiply $-5t \times -7t = 35t^2$

Step 3: Multiply $-5t \times 9 = -45t$

Step 4: Arrange the answers horizontally in order: $-10t^3 + 35t^2 - 45t$

Remove parentheses in the following problems.

1. $3x(3x^2 + 4x - 1)$
2. $4y(y^3 - 7)$
3. $7a^2(2a^2 + 3a + 2)$
4. $-5d^3(d^2 - 5d)$
5. $2w(-4w^2 + 3w - 8)$
6. $8p(p^3 - 6p + 5)$
7. $-9b^2(-2b + 5)$
8. $2t(t^2 - 4t - 10)$
9. $10c(4c^2 + 3c - 7)$
10. $6z(2z^4 - 5z^2 - 4)$
11. $-9t^2(3t^2 + 5t + 6)$
12. $c(-3c - 5)$
13. $3p(p^3 - p^2 - 9)$
14. $-k^2(2k + 4)$
15. $-3(4m^2 - 5m + 8)$
16. $6x(-7x^3 + 10)$
17. $-w(w^2 - 4w + 7)$
18. $2y(5y^2 - y)$
19. $3d(d^5 - 7d^3 + 4)$
20. $-5t(-4t^2 - 8t + 1)$
21. $7(2w^2 - 9w + 4)$
22. $3y^2(y^2 - 11)$
23. $v^2(v^2 + 3v + 3)$
24. $8x(2x^3 + 3x + 1)$
25. $-5d(4d^2 + 7d - 2)$
26. $-k^2(-3k + 6)$
27. $3x(-x^2 - 5x + 5)$
28. $4z(4z^4 - z - 7)$
29. $-5y(9y^3 - 3)$
30. $2b^2(7b^2 + 4b + 4)$

7.6 Dividing Polynomials by Monomials

Example 7: $\dfrac{-8wx + 6x^2 - 16wx^2}{2wx}$

Step 1: Rewrite the problem. Divide each term from the top by the denominator, $2wx$.

$$\frac{-8wx}{2wx} + \frac{6x^2}{2wx} + \frac{-16wx^2}{2wx}$$

Step 2: Simplify each term in the problem. Then combine like terms.

$$-4 + \frac{3x}{w} - 8x$$

Simplify each of the following.

1. $\dfrac{bc^2 - 8bc - 2b^2c^2}{2bc}$

2. $\dfrac{3jk^2 + 12k + 9j^2k}{3jk}$

3. $\dfrac{5x^2y - 8xy^2 + 2y^3}{2xy}$

4. $\dfrac{16st^2 + st - 12s}{4st}$

5. $\dfrac{4wx^2 + 6wx - 12w^3}{2wx}$

6. $\dfrac{cd^2 + 10cd^3 + 16c^2}{2cd}$

7. $\dfrac{y^2z^3 - 2yz - 8z^2}{-2yz^2}$

8. $\dfrac{a^2b + 2ab^2 - 14ab^3}{2a^2}$

9. $\dfrac{pr^2 + 6pr + 8p^2r^2}{2pr^2}$

10. $\dfrac{6xy^2 - 3xy + 18x^2}{-3xy}$

11. $\dfrac{6x^2y + 12xy - 24y^2}{6xy}$

12. $\dfrac{5m^2n - 10mn - 25n^2}{5mn}$

13. $\dfrac{st^2 - 10st - 16s^2t^2}{2st}$

14. $\dfrac{7jk^2 - 14jk - 63j^2}{7jk}$

7.7 Removing Parentheses and Simplifying

In the following problem, you must multiply each set of parentheses by the numbers and variables outside the parentheses, and then add the polynomials to simplify the expressions.

Example 8: $8x\left(2x^2 - 5x + 7\right) - 3x\left(4x^2 + 3x - 8\right)$

Step 1: Multiply to remove the first set of parentheses.

$8x\left(2x^2 - 5x + 7\right) = 16x^3 - 40x^2 + 56x$

Step 2: Multiply to remove the second set of parentheses.

$-3x\left(4x^2 + 3x - 8\right) = -12x^3 - 9x^2 + 24x$

Step 3: Copy each polynomial in columns, making sure the terms with the same variable and exponent are under each other. Add to simplify.

$$\begin{array}{r} 16x^3 - 40x^2 + 56x \\ (+) - 12x^3 - 9x^2 + 24x \\ \hline 4x^3 - 49x^2 + 80x \end{array}$$

Remove the parentheses and simplify the following problems.

1. $4t\left(t + 7\right) + 5t\left(2t^2 - 4t + 1\right)$
2. $-5y\left(3y^2 - 5y + 3\right) - 6y\left(y^2 - 4y - 4\right)$
3. $-3\left(3x^2 + 4x\right) + 5x\left(x^2 + 3x + 2\right)$
4. $2b\left(5b^2 - 8b - 1\right) - 3b\left(4b + 3\right)$
5. $8d^2\left(3d + 4\right) - 7d\left(3d^2 + 4d + 5\right)$
6. $5a\left(3a^2 + 3a + 1\right) - \left(-2a^2 + 5a - 4\right)$
7. $3m\left(m + 7\right) + 8\left(4m^2 + m + 4\right)$
8. $4c^2\left(-6c^2 - 3c + 2\right) - 7c\left(5c^3 + 2c\right)$
9. $-8w\left(-w + 1\right) - 4w\left(3w - 5\right)$
10. $6p\left(2p^2 - 4p - 6\right) + 3p\left(p^2 + 6p + 9\right)$

7.8 Multiplying Two Binomials Using the FOIL Method

When you multiply two binomials such as $(x+6)(x-5)$, you must multiply each term in the first binomial by each term in the second binomial. The easiest way is to use the **FOIL** method. If you can remember the word **FOIL**, it can help you keep order when you multiply. The "**F**" stands for **first**, "**O**" stands for **outside**, "**I**" stands for **inside**, and "**L**" stands for **last**.

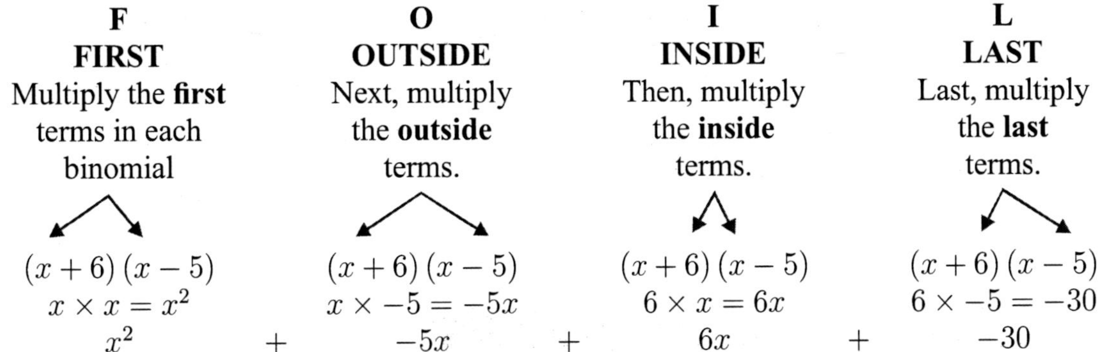

$(x+6)(x-5)$		$(x+6)(x-5)$		$(x+6)(x-5)$		$(x+6)(x-5)$
$x \times x = x^2$		$x \times -5 = -5x$		$6 \times x = 6x$		$6 \times -5 = -30$
x^2	$+$	$-5x$	$+$	$6x$	$+$	-30

Now just combine like terms, $6x - 5x = x$, and write your answer.
$(x+6)(x-5) = x^2 + x - 30$.

Note: It is customary for mathematicians to write polynomials in descending order. That means that the term with the highest exponent comes first in a polynomial. The next highest exponent is second, and so on. When you use the **FOIL** method, the terms will always be in the customary order. You just need to combine like terms and write your answer.

1. $(y-7)(y+3)$
2. $(2x+4)(x+9)$
3. $(4b-3)(3b-4)$
4. $(6g+2)(g-9)$
5. $(7k-5)(-4k-3)$
6. $(8v-2)(3v+4)$
7. $(10p+2)(4p+3)$
8. $(3h-9)(-2h-5)$
9. $(w-4)(w-7)$
10. $(6x+1)(x-2)$
11. $(5t+3)(2t-1)$
12. $(4y-9)(4y+9)$
13. $(a+6)(3a+5)$
14. $(3z-8)(z-4)$
15. $(5c+2)(6c+5)$
16. $(y+3)(y-3)$
17. $(2w-5)(4w+6)$
18. $(7x+1)(x-4)$
19. $(6t-9)(4t-4)$
20. $(5b+6)(6b+2)$
21. $(2z+1)(10z+4)$
22. $(11w-8)(w+3)$
23. $(5d-9)(9d+9)$
24. $(9g+2)(g-2)$
25. $(4p+7)(2p+3)$
26. $(m+5)(m-5)$
27. $(8b-8)(2b-1)$
28. $(z+3)(3z+5)$
29. $(7y-5)(y-3)$
30. $(9x+5)(3x-1)$
31. $(3t+1)(t+10)$
32. $(2w-9)(8w+7)$
33. $(8s-2)(s+4)$
34. $(4k-1)(8k+9)$
35. $(h+12)(h-2)$
36. $(3x+7)(7x+3)$
37. $(2v-6)(2v+6)$
38. $(2x+8)(2x-3)$
39. $(k-1)(6k+12)$
40. $(3w+11)(2w+2)$
41. $(8y-10)(5y-3)$
42. $(6d+13)(d-1)$
43. $(7h+3)(2h+4)$
44. $(5n+9)(5n-5)$
45. $(6z+5)(z-8)$

7.9 Simplifying Expressions with Exponents

Example 9: **Simplify** $(2a+5)^2$

When you simplify an expression such as $(2a+5)^2$, write the expression as two binomials and use FOIL to simplify.

$(2a+5)^2 = (2a+5)(2a+5)$

Using FOIL we have $4a^2 + 10a + 10a + 25 = 4a^2 + 20a + 25$

Example 10: **Simplify** $4(3a+2)^2$

Using order of operations, we must simplify the exponent first.

$4(3a+2)^2$

$4(3a+2)(3a+2)$

$4(9a^2 + 6a + 6a + 4)$

$4(9a^2 + 12a + 4)$ Now multiply by 4.

$4(9a^2 + 12a + 4) = 36a^2 + 48a + 16$

Multiply the following binomials.

1. $(y+3)^2$
2. $2(2x+4)^2$
3. $6(4b-3)^2$
4. $5(6g+2)^2$
5. $(-4k-3)^2$
6. $3(-2h-5)^2$
7. $-2(8v-2)^2$
8. $(10p+2)^2$
9. $6(-2h-5)^2$
10. $6(w-7)^2$
11. $2(6x+1)^2$
12. $(9x+2)^2$
13. $(5t+3)^2$
14. $3(4y-9)^2$
15. $8(a+6)^2$
16. $4(3z-8)^2$
17. $3(5c+2)^2$
18. $4(3t+9)^2$

7.10 Multiplying Binomials by Trinomials

When multiplying a binomial and a trinomial together, the new expression is a **cubic expression**. A cubic expression is one in which the variable is raised to the third power, such as x^3. This cube is the largest exponent in the expression. The same variable will most likely be raised to the second and first powers also, depending on the coefficient of the variable.

Example 11: Simplify: $(x-2)(x^2+x-6)$

Step 1: Similarly to the FOIL method, every term in the binomial, $(x-2)$, must be multiplied by every term in the trinomial, (x^2+x-6).
The first binomial term, x, is multiplied by each term in the trinomial.

$$\begin{aligned} x \times x^2 &= x^3 \\ x \times x &= x^2 \\ x \times -6 &= -6x \end{aligned}$$

The second binomial term, -2, is multiplied by each term in the trinomial.

$$\begin{aligned} -2 \times x^2 &= -2x^2 \\ -2 \times x &= -2x \\ -2 \times -6 &= 12 \end{aligned}$$

Step 2: Add all of the new terms together. $x^3+x^2-6x-2x^2-2x+12$

Step 3: Combine like terms and simplify. Be sure to write the expression in descending order with respect to the exponent of the variable. This means the term that contains the variable with the largest exponent, such as x^3, is first in the expression. The constant is last because the constant does not have a variable, it is just a number.

$$\begin{aligned} x^3+x^2-6x-2x^2-2x+12 &= x^3+x^2-2x^2-6x-2x+12 \\ &= x^3-x^2-8x+12 \end{aligned}$$

The simplified answer is $x^3-x^2-8x+12$.

Simplify the following by multiplying the binomials and trinomials together.

1. $(x+1)(x^2-2x-3)$
2. $(x-5)(x^2+2x-1)$
3. $(x+3)(x^2-7x+4)$
4. $(x+2)(x^2+4x+5)$
5. $(x-3)(x^2+9x-8)$
6. $(x+6)(x^2-2x+6)$
7. $(x+5)(x^2+x+1)$
8. $(x-4)(x^2-x-5)$
9. $(x-1)(x^2+3x+3)$
10. $(x-7)(x^2+2x+1)$
11. $(x+10)(x^2-4x+2)$
12. $(x-9)(x^2-9x+9)$
13. $(x+2)(x^2+4x-7)$
14. $(x+5)(x^2-x-2)$
15. $(x-8)(x^2-10x+30)$

Chapter 7 Review

Simplify.

1. $3a^2 + 9a^2$
2. $(7x^2y^4)(9xy^5)$
3. $-6z^2(z+3)$
4. $(4b^2)(5b^3)$
5. $7x^2 - 9x^2$
6. $(5p-4)-(3p+2)$
7. $-5t(3t+9)^2$
8. $(3w^3y^2)(4wy^5)$
9. $3(2g+3)^2$
10. $14d^4 - 9d^4$
11. $(7w-4)(w-8)$
12. $(9x+2)(x+5)$
13. $4y(4y^2-9y+2)$
14. $(8a^4b)(2ab^3)(ab)$
15. $(5w^6)(9w^9)$
16. $8w^3 + 12x^3$
17. $15p^5 - 11p^5$
18. $(3s^4t^2)(4st^3)$
19. $(4d+9)(2d+7)$
20. $4w(-3w^2+7w-5)$
21. $24z^6 - 10z^6$
22. $-7y^3 - 8y^3$
23. $(a^2v)(2av)(a^3v^6)$
24. $4(6y-5)^2$
25. $(4x^5y^3)(2xy^3)$
26. $24z^6 - 10z^6$
27. $(3p^3-1)(p+5)$
28. $2b(b-4)-(b^2+2b+1)$
29. $(6k^2+5k)+(k^2+k+9)$
30. $(q^2r^3)(3qr^2)(2q^4r)$
31. $(x-1)(x^2+2x-6)$
32. $(1-y)(y^2+4y+5)$
33. $(x+3)(x^2-3x+9)$

Chapter 8
Factoring

This chapter covers the following North Carolina mathematics standards for Algebra I:

Competency Goal	Objectives
Number and Operations	1.01b 1.01c

8.1 Finding the Greatest Common Factor of Polynomials

In a multiplication problem, the numbers multiplied together are called **factors**. The answer to a multiplication problem is a called the **product**.

In the multiplication problem $5 \times 4 = 20$, 5 and 4 are factors and 20 is the product.

If we reverse the problem, $20 = 5 \times 4$, we say we have **factored** 20 into 5×4.

In this chapter, we will factor **polynomials**.

Example 1: Factor by finding the greatest common factor of $2y^3 + 6y^2$.

Step 1: Look at the whole numbers. The greatest common factor of 2 and 6 is 2. Factor the 2 out of each term.
$2\left(y^3 + 3y^2\right)$

Step 2: Look at the remaining terms, $y^3 + 3y^2$. What are the common factors of each term?

$$\begin{array}{ccccc} y^3 & = & y & \times & \boxed{y \times y} \\ 3y^2 & = & 3 & \times & \boxed{y \times y} \end{array} \longleftarrow \text{common factors} = y^2$$

Step 3: Factor 2 and y^2 out of each term: $2y^2\left(y+3\right)$

Check: $2y^2\left(y+3\right) = 2y^3 + 6y^2$

Factor by finding the greatest common factor of each of the following.

1. $6x^4 + 18x^2$
2. $14y^3 + 7y$
3. $4b^5 + 12b^3$
4. $10a^3 + 5$
5. $2y^3 + 8y^2$
6. $6x^4 - 12x^2$
7. $18y^2 - 12y$
8. $15a^3 - 25a^2$
9. $4x^3 + 16x^2$
10. $6b^2 + 21b^5$
11. $27m^3 + 18m^4$
12. $100x^4 - 25x^3$
13. $4b^4 - 12b^3$
14. $18c^2 + 24c$
15. $20y^3 + 30y^5$
16. $16x^2 - 24x^5$
17. $15a^4 - 25a^2$
18. $24b^3 + 16b^6$
19. $36y^4 + 9y^2$
20. $42x^3 + 49x$

Factoring larger polynomials with 3 or 4 terms works the same way.

Example 2: $4x^5 + 16x^4 + 12x^3 + 8x^2$

Step 1: Find the greatest common factor of the whole numbers. 4 can be divided evenly into 4, 16, 12, and 8; therefore, 4 is the greatest common factor.

Step 2: Find the greatest common factor of the variables. x^5, x^4, x^3, and x^2 can be divided by x^2, the lowest power of x in each term.

$$4x^5 + 16x^4 + 12x^3 + 8x^2 = 4x^2\left(x^3 + 4x^2 + 3x + 2\right)$$

Factor each of the following polynomials.

1. $5a^3 + 15a^2 + 20a$
2. $18y^4 + 6y^3 + 24y^2$
3. $12x^5 + 21x^3 + x^2$
4. $6b^4 + 3b^3 + 15b^2$
5. $14c^3 + 28c^2 + 7c$
6. $15b^4 - 5b^2 + 20b$
7. $t^3 + 3t^2 - 5t$
8. $8a^3 - 4a^2 + 12a$
9. $16b^5 - 12b^4 - 10b^2$
10. $20x^4 + 16x^3 - 24x^2 + 28x$
11. $40b^7 + 30b^5 - 50b^3$
12. $20y^4 - 15y^3 + 30y^2$
13. $4m^5 + 8m^4 + 12m^3 + 6m^2$
14. $16x^5 + 20x^4 - 12x^3 + 24x^2$
15. $18y^4 + 21y^3 - 9y^2$
16. $3n^5 + 9n^3 + 12n^2 + 15n$
17. $4d^6 - 8d^2 + 2d$
18. $10w^2 + 4w + 2$
19. $6t^3 - 3t^2 + 9t$
20. $25p^5 - 10p^3 - 5p^2$
21. $18x^4 + 9x^2 - 36x$
22. $6b^4 - 12b^2 - 6b$
23. $y^3 + 3y^2 - 9y$
24. $10x^5 - 2x^4 + 4x^2$

Example 3: Find the greatest common factor of $4a^3b^2 - 6a^2b^2 + 2a^4b^3$

Step 1: The greatest common factor of the whole numbers is 2.

$4a^3b^2 - 6a^2b^2 + 2a^4b^3 = 2\left(2a^3b^2 - 3a^2b^2 + a^4b^3\right)$

Step 2: Find the lowest power of each variable that is in each term. Factor them out of each term. The lowest power of a is a^2. The lowest power of b is b^2.

$4a^3b^2 - 6a^2b^2 + 2a^4b^3 = 2a^2b^2\left(2a - 3 + a^2b\right)$

Factor each of the following polynomials.

1. $3a^2b^2 - 6a^3b^4 + 9a^2b^3$
2. $12x^4y^3 + 18x^3y^4 - 24x^3y^3$
3. $20x^2y - 25x^3y^3$
4. $12x^2y - 20x^2y^2 + 16xy^2$
5. $8a^3b + 12a^2b + 20a^2b^3$
6. $36c^4 + 42c^3 + 24c^2 - 18c$
7. $14m^3n^4 - 28m^3n^2 + 42m^2n^3$
8. $16x^4y^2 - 24x^3y^2 + 12x^2y^2 - 8xy^2$
9. $32c^3d^4 - 56c^2d^3 + 64c^3d^2$
10. $21a^4b^3 + 27a^2b^3 + 15a^3b^2$
11. $4w^3t^2 + 6w^2t - 8wt^2$
12. $5pw^3 - 2p^2q^2 - 9p^3q$
13. $49x^3t^3 + 7xt^2 - 14xt^3$
14. $9cd^4 - 3d^4 - 6c^2d^3$
15. $12a^2b^3 - 14ab + 10ab^2$
16. $25x^4 + 10x - 20x^2$
17. $bx^3 - b^2x^2 + b^3x$
18. $4k^3a^2 + 22ka + 16k^2a^2$
19. $33w^4y^2 - 9w^3y^2 + 24w^2y^2$
20. $18x^3 - 9x^5 + 27x^2$

8.2 Factor by Grouping

Not all polynomials have a common factor in each term. In this case they may sometimes be factored by grouping.

Example 4: Factor $ab + 4a + 2b + 8$

Step 1: Factor an a from the first two terms and a 2 from the last two terms.

$a\,(b + 4) + 2\,(b + 4)$

Now the polynomial has two terms, $a\,(b + 4)$ and $2\,(b + 4)$. Notice that $(b + 4)$ is a factor of each term.

Step 2: Factor out the common factor of each term:

$ab + 4a + 2b + 8 = (b + 4)\,(a + 2).$

Check: Multiply using the FOIL method to check.

$(b + 4)\,(a + 2) = ab + 4a + 2b + 8$

Factor the following polynomials by grouping.

1. $xy + 4x + 2y + 8$
2. $cd + 5c + 4d + 20$
3. $xy - 4x + 6y - 24$
4. $ab + 6a + 3b + 18$
5. $ab + 3a - 5b - 15$
6. $xy - 2x + 6x - 12$
7. $cd + 4c + 4d + 16$
8. $mn - 5m + 3n - 15$
9. $ab + 4a + 3b + 12$
10. $xy + 7x - 4y - 28$
11. $ab - 2a + 8b - 16$
12. $cd + 4c - 5d - 20$
13. $mn + 6m - 2n - 12$
14. $xy - 9x - 3y + 27$
15. $bc - 3b + 5c - 15$
16. $ab + a + 7b + 7$
17. $xy + 4y + 2y + 8$
18. $cd + 9c - d - 9$
19. $ab + 2a - 7b - 14$
20. $xy - 6x - 2y + 12$
21. $wz + 6z - 4w - 24$

8.3 Factoring Trinomials

In the chapter on polynomials, you multiplied binomials (two terms) together, and the answer was a trinomial (three terms).

For example, $(x+6)(x-5) = x^2 + x - 30$

Now, you need to practice factoring a trinomial into two binomials.

Example 5: Factor $x^2 + 6x + 8$

Step 1: When the trinomial is in descending order as in the example above, you need to find a pair of numbers whose sum equals the number in the second term, while their product equals the third term. In the above example, find the pair of numbers that has a sum of 6 and a product of 8.

_____ + _____ = 6 and _____ × _____ = 8

The pair of numbers that satisfies both equations is 4 and 2.

Step 2: Use the pair of numbers in the binomials.

The factors of $x^2 + 6x + 8$ are $(x+4)(x+2)$

Check: To check, use the FOIL method.
$(x+4)(x+2) = x^2 + 4x + 2x + 8 = x^2 + 6x + 8$

Notice: when the second term and the third term of the trinomial are both positive, both numbers in the solution are positive.

Example 6: Factor $x^2 - x - 6$ Find the pair of numbers where:

the sum is -1 and the product is -6

_____ + _____ = -1 and _____ × _____ = -6

The pair of numbers that satisfies both equations is 2 and -3.
The factors of $x^2 - x - 6$ are $(x+2)(x-3)$

Notice, if the second term and the third term are negative, one number in the solution pair is positive, and the other number is negative.

Example 7: Factor $x^2 - 7x + 12$ Find the pair of numbers where:

the sum is -7 and the product is 12

_____ + _____ = -7 and _____ × _____ = 12

The pair of numbers that satisfies both equations is -3 and -4
The factors of $x^2 - 7x + 12$ are $(x-3)(x-4)$.

Notice, if the second term of a trinomial is negative and the third term is positive, both numbers in the solution are negative.

Find the factors of the following trinomials.

1. $x^2 - x - 2$
2. $y^2 + y - 6$
3. $w^2 + 3w - 4$
4. $t^2 + 5t + 6$
5. $x^2 + 2x - 8$
6. $k^2 - 4k + 3$
7. $t^2 + 3t - 10$
8. $x^2 - 3x - 4$
9. $y^2 - 5y + 6$
10. $y^2 + y - 20$
11. $a^2 - a - 6$
12. $b^2 - 4b - 5$
13. $c^2 - 5c - 14$
14. $c^2 - c - 12$
15. $d^2 + d - 6$
16. $x^2 - 3x - 28$
17. $y^2 + 3y - 18$
18. $a^2 - 9a + 20$
19. $b^2 - 2b - 15$
20. $c^2 + 7c - 8$
21. $t^2 - 11t + 30$
22. $w^2 + 13w + 36$
23. $m^2 - 2m - 48$
24. $y^2 + 14y + 49$
25. $x^2 + 7x + 10$
26. $a^2 - 7a + 6$
27. $d^2 - 6d - 27$

8.4 More Factoring Trinomials

Sometimes a trinomial has a greatest common factor which must be factored out first.

Example 8: Factor $4x^2 + 8x - 32$

Step 1: Begin by factoring out the greatest common factor, 4.
$4\left(x^2 + 2x - 8\right)$

Step 2: Factor by finding a pair of numbers whose sum is 2 and product is -8.
4 and -2 will work, so
$4\left(x^2 + 2x - 8\right) = 4\left(x + 4\right)\left(x - 2\right)$

Check: Multiply to check. $4\left(x + 4\right)\left(x - 2\right) = 4x^2 + 8x - 32$

Factor the following trinomials. Be sure to factor out the greatest common factor first.

1. $2x^2 + 6x + 4$
2. $3y^2 - 9y + 6$
3. $2a^2 + 2a - 12$
4. $4b^2 + 28b + 40$
5. $3y^2 - 6y - 9$
6. $10x^2 + 10x - 200$
7. $5c^2 - 10c - 40$
8. $6d^2 + 30d - 36$
9. $4x^2 + 8x - 60$
10. $6a^2 - 18a - 24$
11. $5b^2 + 40b + 75$
12. $3c^2 - 6c - 24$
13. $2x^2 - 18x + 28$
14. $4y^2 - 20y + 16$
15. $7a^2 - 7a - 42$
16. $6b^2 - 18b - 60$
17. $11d^2 + 66d + 88$
18. $3x^2 - 24x + 45$

8.5 Factoring More Trinomials

Some trinomials have a whole number in front of the first term that cannot be factored out of the trinomial. The trinomial can still be factored.

Example 9: Factor $2x^2 + 5x - 3$

Step 1: To get a product of $2x^2$, one factor must begin with $2x$ and the other with x.

$(2x \quad)(x \quad)$

Step 2: Now think: What two numbers give a product of -3? The two possibilities are 3 and -1 or -3 and 1. We know they could be in any order so there are 4 possible arrangements.

$(2x + 3)(x - 1)$
$(2x - 3)(x + 1)$
$(2x + 1)(x - 3)$
$(2x - 1)(x + 3)$

Step 3: Multiply each possible answer until you find the arrangement of numbers that works. Multiply the outside terms and the inside terms and add them together to see which one will equal $5x$.

$(2x + 3)(x - 1) = 2x^2 + x - 3$
$(2x - 3)(x + 1) = 2x^2 - x - 3$
$(2x + 1)(x - 3) = 2x^2 - 5 - 3$
$\boxed{(2x - 1)(x + 3) = 2x^2 + 5x - 3} \longleftarrow$ This arrangement works, therefore:

The factors of $2x^2 + 5x - 3$ are $(2x - 1)(x + 3)$

Factor the following trinomials.

1. $3y^2 + 14y + 8$
2. $5a^2 + 24a - 5$
3. $7b^2 + 30b + 8$
4. $2c^2 - 9c + 9$
5. $2y^2 - 7y - 15$
6. $3x^2 + 4x + 1$
7. $7y^2 + 13y - 2$
8. $11a^2 + 35a + 6$
9. $5y^2 + 17y - 12$
10. $3a^2 + 4a - 7$
11. $2a^2 + 3a - 20$
12. $5b^2 - 13b - 6$
13. $3y^2 - 17x + 36$
14. $2x^2 - 17x + 36$
15. $11x^2 - 29x - 12$
16. $5c^2 + 2c - 16$
17. $7y^2 - 30y + 27$
18. $2x^2 - 3x - 20$
19. $5b^2 + 24b - 5$
20. $7d^2 + 18d + 8$
21. $3x^2 - 20x + 25$
22. $2a^2 - 7a - 4$
23. $5m^2 + 12m + 4$
24. $9y^2 - 5y - 4$
25. $2b^2 - 13b + 18$
26. $7x^2 + 31x - 20$
27. $3c^2 - 2c - 21$

8.6 Factoring Trinomials with Two Variables

Some trinomials have two variables with exponents. You can still factor these trinomials.

Example 10: Factor $x^2 + 5xy + 6y^2$

Step 1: Notice there is an x^2 in the first term and a y^2 in the last term. When you see two different terms that are squared, you know there has to be an x and a y in each factor:

$(x \quad y)(x \quad y)$

Step 2: Now think: What are two numbers whose sum is 5 and product is 6? You see that 3 and 2 will work. Put 3 and 2 in the factors:

$(x + 3y)(x + 2y)$

Check: Multiply to check. $(x + 3y)(x + 2y) = x^2 + 3xy + 2xy + 6y^2 = x^2 + 5xy + 6y^2$

Factor the following trinomials.

1. $a^2 + 6ab + 8b^2$
2. $x^2 + 3xy - 4y^2$
3. $c^2 - 2cd - 15d^2$
4. $g^2 + 7gh + 10h^2$
5. $a^2 - 5ab + 6b^2$
6. $c^2 - cd - 30d^2$
7. $x^2 + 5xy - 24y^2$
8. $a^2 - 4ab + 4b^2$
9. $c^2 - 11cd + 30d^2$
10. $x^2 - 6xy + 8y^2$
11. $g^2 - gh - 42h^2$
12. $a^2 - ab - 20b^2$
13. $x^2 + 12xy + 32y^2$
14. $c^2 + 3cd - 40d^2$
15. $x^2 + 6xy - 27y^2$
16. $a^2 - 2ab - 48b^2$
17. $c^2 - 3cd - 28d^2$
18. $x^2 + xy - 6y^2$

8.7 Factoring the Difference of Two Squares

The product of a term and itself is called a **perfect square**.

25 is a perfect square because $5 \times 5 = 25$
49 is a perfect square because $7 \times 7 = 49$

Any variable with an even exponent is a perfect square.

y^2 is a perfect square because $y \times y = y^2$
y^4 is a perfect square because $y^2 \times y^2 = y^4$

When two terms that are both perfect squares are subtracted, factoring those terms is very easy. To factor the difference of perfect squares, you use the square root of each term, a plus sign in the first factor, and a minus sign in the second factor.

Example 11: Factor $4x^2 - 9$

This example has two terms which are both perfect squares, and the terms are subtracted.

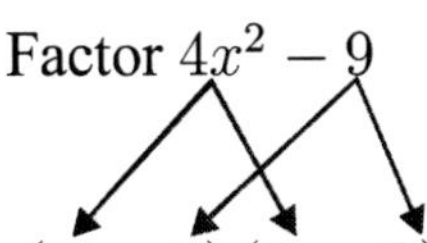
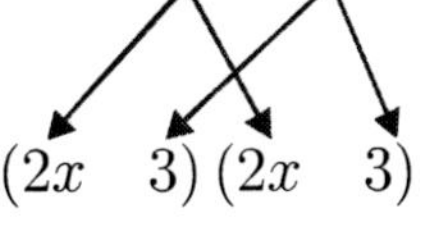

Step 1: $(2x \quad 3)\,(2x \quad 3)$

Find the square root of each term. Use the square roots in each of the factors.

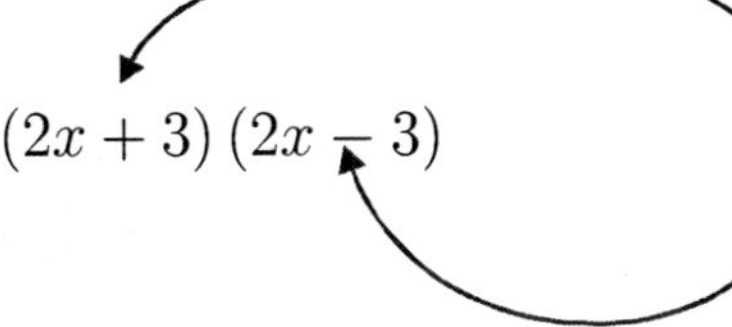

Step 2: $(2x + 3)\,(2x - 3)$

Use a plus sign in one factor and a minus sign in the other factor.

Check: Multiply to check. $(2x + 3)\,(2x - 3) = 4x^2 - 6x + 6x - 9 = 4x^2 - 9$

The inner and outer terms add to zero.

Example 12: Factor $81y^4 - 1$

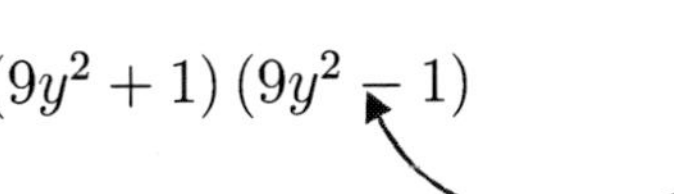

Step 1: $(9y^2 + 1)\,(9y^2 - 1)$

Factor like the example above. Notice, the second factor is also the difference of two perfect squares.

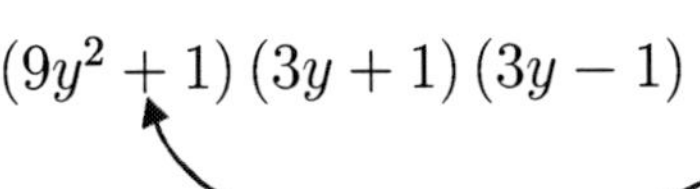

Step 2: $(9y^2 + 1)\,(3y + 1)\,(3y - 1)$

Factor the second term further. **Note: You cannot factor the sum of two perfect squares.**

Check: Multiply in reverse to check your answer.
$(9y^2 + 1)\,(3y + 1)\,(3y - 1) = (9y^2 + 1)\,(9y^2 - 3y + 3y - 1) =$
$(9y^2 + 1)\,(9y^2 - 1) = 81y^4 + 9y^2 - 9y^2 - 1 = 81y^4 - 1$

Factor the following differences of perfect squares.

1. $64x^2 - 49$
2. $4y^4 - 25$
3. $9a^4 - 4$
4. $25c^4 - 9$
5. $64y^2 - 9$
6. $x^4 - 16$
7. $49x^2 - 4$
8. $4d^2 - 25$
9. $9a^2 - 16$
10. $100y^4 - 49$
11. $c^4 - 36$
12. $36x^2 - 25$
13. $25x^2 - 4$
14. $9x^4 - 64$
15. $49x^2 - 100$
16. $16x^2 - 81$
17. $9y^4 - 1$
18. $64c^2 - 25$
19. $25d^2 - 64$
20. $36a^4 - 49$
21. $16x^4 - 16$
22. $b^2 - 25$
23. $c^4 - 144$
24. $9y^2 - 4$
25. $81x^4 - 16$
26. $4b^2 - 36$
27. $9w^2 - 9$
28. $64a^2 - 25$
29. $49y^2 - 121$
30. $x^6 - 9$

Chapter 8 Review

Factor the following polynomials completely.

1. $8x - 18$
2. $6x^2 - 18x$
3. $16b^3 + 8b$
4. $15a^3 + 40$
5. $20y^6 - 12y^4$
6. $5a - 15a^2$
7. $4y^2 - 36$
8. $25a^4 - 49b^2$
9. $3ax + 3ay + 4x + 4y$
10. $ax - 2x + ay - 2y$
11. $2bx + 2x - 2by - 2y$
12. $2b^2 - 2b - 12$
13. $yx^3 + 14x - 3x^2 - 6$
14. $3a^3 + 4a^2 + 9a + 12$
15. $27y^2 + 42y - 5$
16. $12b^2 + 25b - 7$
17. $c^2 + cd - 20d^2$
18. $x^2 - 4xy - 21y^2$
19. $6y^2 + 30y + 36$
20. $2b^2 + 6b - 20$
21. $16b^4 - 81d^4$
22. $9w^2 - 54w - 63$
23. $m^2p^2 - 5mp + 2m^2p - 10m$
24. $12x^2 + 27x$
25. $2xy - 36 + 8y - 9x$
26. $2a^4 - 32$
27. $21c^2 + 41c + 10$
28. $x^2 - y + xy - x$
29. $2b^3 - 24 + 16b - 3b^2$
30. $5 - 2a - 25a^2 + 10a^3$

Chapter 9
Solving Quadratic Equations

This chapter covers the following North Carolina mathematics standards for Algebra I:

Competency Goal	Objectives
Algebra	4.02

In the previous chapter, we factored polynomials such as $y^2 - 4y - 5$ into two factors:

$$y^2 - 4y - 5 = (y+1)(y-5)$$

In this chapter, we learn that any equation that can be put in the form $ax^2+bx+c=0$ is a quadratic equation if a, b, and c are real numbers and $a \neq 0$. $ax^2+bx+c=0$ is the standard form of a quadratic equation. To solve these equations, follow the steps below.

Example 1: Solve $y^2 - 4y - 5 = 0$

Step 1: Factor the left side of the equation.

$$\begin{aligned} y^2 - 4y - 5 &= 0 \\ (y+1)(y-5) &= 0 \end{aligned}$$

Step 2: If the product of these two factors equals zero, then the two factors individually must be equal to zero. Therefore, to solve, we set each factor equal to zero.

$$\begin{array}{rr} (y+1) & =0 \\ -1 & -1 \\ \hline y & =-1 \end{array} \qquad \begin{array}{rr} (y-5) & =0 \\ +5 & +5 \\ \hline y & =5 \end{array}$$

The equation has two solutions: $y = -1$ and $y = 5$

Check: To check, substitute each solution into the original equation.

When $y = -1$, the equation becomes:
$$\begin{aligned} (-1)^2 - (4)(-1) - 5 &= 0 \\ 1 + 4 - 5 &= 0 \\ 0 &= 0 \end{aligned}$$

When $y = 5$, the equation becomes:
$$\begin{aligned} 5^2 - (4)(5) - 5 &= 0 \\ 25 - 20 - 5 &= 0 \\ 0 &= 0 \end{aligned}$$

Both solutions produce true statements.
The solution set for the equation is $\{-1, 5\}$

Solve each of the following quadratic equations by factoring and setting each factor equal to zero. Check by substituting answers back in the original equation.

1. $x^2 + x - 6 = 0$
2. $y^2 - 2y - 8 = 0$
3. $a^2 + 2a - 15 = 0$
4. $y^2 - 5y + 4 = 0$
5. $b^2 - 9b + 14 = 0$
6. $x^2 - 3x - 4 = 0$
7. $y^2 + y - 20 = 0$
8. $d^2 + 6d + 8 = 0$
9. $y^2 - 7y + 12 = 0$
10. $x^2 - 3x - 28 = 0$
11. $a^2 - 5a + 6 = 0$
12. $b^2 + 3b - 10 = 0$
13. $a^2 + 7a - 8 = 0$
14. $c^2 + 3x + 2 = 0$
15. $x^2 - x - 42 = 0$
16. $a^2 + a - 6 = 0$
17. $b^2 + 7b + 12 = 0$
18. $y^2 + 2y - 15 = 0$
19. $a^2 - 3a - 10 = 0$
20. $d^2 + 10d + 16 = 0$
21. $x^2 - 4x - 12 = 0$

Quadratic equations that have a whole number and a variable in the first term are solved the same way as the previous page. Factor the trinomial, and set each factor equal to zero to find the solution set.

Example 2: Solve $2x^2 + 3x - 2 = 0$
$(2x - 1)(x + 2) = 0$
Set each factor equal to zero and solve:

$$\begin{array}{rr} 2x - 1 & = 0 \\ +1 & +1 \\ \hline \frac{2x}{2} & = \frac{1}{2} \\ x & = \frac{1}{2} \end{array} \qquad \begin{array}{rr} x + 2 & = 0 \\ -2 & -2 \\ \hline x & = -2 \end{array}$$

The solution set is $\left\{\frac{1}{2}, -2\right\}$.

Solve the following quadratic equations.

22. $3y^2 + 4y - 32 = 0$
23. $5c^2 - 2c - 16 = 0$
24. $7d^2 + 18d + 8 = 0$
25. $3a^2 - 10a - 8 = 0$
26. $11x^2 - 31x - 6 = 0$
27. $5b^2 + 17b + 6 = 0$
28. $3x^2 - 11x - 20 = 0$
29. $5a^2 + 47a - 30 = 0$
30. $2c^2 - 5c - 25 = 0$
31. $2y^2 + 11y - 21 = 0$
32. $5a^2 + 23a - 42 = 0$
33. $3d^2 + 11d - 20 = 0$
34. $3x^2 - 10x + 8 = 0$
35. $7b^2 + 23b - 20 = 0$
36. $9a^2 - 58a + 24 = 0$
37. $4c^2 - 25c - 21 = 0$
38. $8d^2 + 53d + 30 = 0$
39. $4y^2 + 37a - 15 = 0$
40. $8a^2 + 37a - 15 = 0$
41. $3x^2 - 41x + 26 = 0$
42. $8b^2 + 2b - 3 = 0$

9.1 Solving the Difference of Two Squares

To solve the difference of two squares, first factor. Then set each factor equal to zero.

Example 3: $25x^2 - 36 = 0$

Step 1: Factor the left side of the equation.

$25x^2 - 36 = 0$
$(5x + 6)(5x - 6) = 0$

Step 2: Set each factor equal to zero and solve.

$$\begin{array}{rl} 5x + 6 & = 0 \\ -6 & -6 \\ \hline \frac{5x}{5} & = \frac{6}{5} \\ x & = -\frac{6}{5} \end{array} \qquad \begin{array}{rl} 5x - 6 & = 0 \\ +6 & +6 \\ \hline \frac{5x}{5} & = \frac{6}{5} \\ x & = \frac{6}{5} \end{array}$$

Check: Substitute each solution in the equation to check.

for $x = -\frac{6}{5}$:

$25x^2 - 36 = 0$

$25\left(-\frac{6}{5}\right)\left(-\frac{6}{5}\right) - 36 = 0 \longleftarrow$ Substitute $-\frac{6}{5}$ for x.

$25\left(\frac{36}{25}\right) - 36 = 0 \longleftarrow$ Cancel the 25's.

$36 - 36 = 0 \longleftarrow$ A true statement. $x = -\frac{6}{5}$ is a solution.

for $x = \frac{6}{5}$:

$25x^2 - 36 = 0$

$25\left(\frac{6}{5}\right)\left(\frac{6}{5}\right) - 36 = 0 \longleftarrow$ Substitute $\frac{6}{5}$ for x.

$25\left(\frac{36}{25}\right) - 36 = 0 \longleftarrow$ Cancel the 25's.

$36 - 36 = 0 \longleftarrow$ A true statement. $x = \frac{6}{5}$ is a solution.

The solution set is $\left\{-\frac{6}{5}, \frac{6}{5}\right\}$.

Find the solution sets for the following.

1. $25a^2 - 16 = 0$

2. $c^2 - 36 = 0$

3. $9x^2 - 64 = 0$

4. $100y^2 - 49 - 0$

5. $4b^2 - 81 = 0$

6. $d^2 - 25 = 0$

7. $9x^2 - 1 = 0$

8. $16a^2 - 9 = 0$

9. $36y^2 - 1 = 0$

10. $36y^2 - 25 = 0$

11. $d^2 - 16 = 0$

12. $64b^2 - 9 = 0$

13. $81a^2 - 4 = 0$

14. $64y^2 - 25 = 0$

15. $4c^2 - 49 = 0$

16. $x^2 - 81 = 0$

17. $49b^2 - 9 = 0$

18. $a^2 - 64 = 0$

19. $9x^2 - 1 = 0$

20. $4y^2 - 9 = 0$

21. $t^2 - 100 = 0$

22. $16k^2 - 81 = 0$

23. $81a^2 - 4 = 0$

24. $36b^2 - 16 = 0$

9.2 Solving Perfect Squares

When the square root of a constant, variable, or polynomial results in a constant, variable, or polynomial without irrational numbers, the expression is a **perfect square**. Some examples are 49, x^2, and $(x-2)^2$.

Example 4: Solve the perfect square for x. $(x-5)^2 = 0$

Step 1: Take the square root of both sides.
$\sqrt{(x-5)^2} = \sqrt{0}$
$(x-5) = 0$

Step 2: Solve the equation.
$(x-5) = 0$
$x - 5 + 5 = 0 + 5$
$x = 5$

Example 5: Solve the perfect square for x. $(x-5)^2 = 64$

Step 1: Take the square root of both sides.
$\sqrt{(x-5)^2} = \sqrt{64}$
$(x-5) = \pm 8$
$(x-5) = 8$ and $(x-5) = -8$

Step 2: Solve the two equations.

$(x-5) = 8$	and	$(x-5) = -8$
$x - 5 + 5 = 8 + 5$	and	$x - 5 + 5 = -8 + 5$
$x = 13$	and	$x = -3$

Solve the perfect square for x.

1. $(x-5)^2 = 0$
2. $(x+1)^2 = 0$
3. $(x+11)^2 = 0$
4. $(x-4)^2 = 0$
5. $(x-1)^2 = 0$
6. $(x+8)^2 = 0$
7. $(x+3)^2 = 4$
8. $(x-5)^2 = 16$
9. $(x-10)^2 = 100$
10. $(x+9)^2 = 9$
11. $(x-4.5)^2 = 25$
12. $(x+7)^2 = 36$
13. $(x+2)^2 = 49$
14. $(x-1)^2 = 4$
15. $(x+8.9)^2 = 49$
16. $(x-6)^2 = 81$
17. $(x-12)^2 = 121$
18. $(x+2.5)^2 = 64$

9.3 Completing the Square

"Completing the Square" is another way of factoring a quadratic equation. To complete the square, convert the equation into a perfect square.

Example 6: Solve $x^2 - 10x + 9 = 0$ by completing the square.

Completing the square:

Step 1: The first step is to get the constant to the other side of the equation. Subtract 9 from both sides:
$x^2 - 10x + 9 - 9 = -9$
$x^2 - 10x = -9$

Step 2: Determine the coefficient of the x. The coefficient in this example is -10. Divide the coefficient by 2 and square the result.
$(-10 \div 2)^2 = (-5)^2 = 25$

Step 3: Add the resulting value, 25, to both sides:
$x^2 - 10x + 25 = -9 + 25$
$x^2 - 10x + 25 = 16$

Step 4: Now factor the $x^2 - 10x + 25$ into a perfect square:
$(x - 5)^2 = 16$

Solving the perfect square:

Step 5: Take the square root of both sides.
$\sqrt{(x-5)^2} = \sqrt{16}$
$(x - 5) = \pm 4$
$(x - 5) = 4$ and $(x - 5) = -4$

Step 6: Solve the two equations.

$(x - 5) = 4$	and	$(x - 5) = -4$
$x - 5 + 5 = 4 + 5$	and	$x - 5 + 5 = -4 + 5$
$x = 9$	and	$x = 1$

Solve for x by completing the square.

1. $x^2 + 2x - 3 = 0$
2. $x^2 - 8x + 7 = 0$
3. $x^2 + 6x - 7 = 0$
4. $x^2 - 16x - 36 = 0$
5. $x^2 - 14x + 49 = 0$
6. $x^2 - 4x = 0$
7. $x^2 + 12x + 27 = 0$
8. $x^2 + 2x - 24 = 0$
9. $x^2 + 12x - 85 = 0$
10. $x^2 - 8x + 15 = 0$
11. $x^2 - 16x + 60 = 0$
12. $x^2 - 8x - 48 = 0$

Chapter 9 Review

Factor and solve each of the following quadratic equations.

1. $16b^2 - 25 = 0$

2. $a^2 - a - 30 = 0$

3. $x^2 - x = 6$

4. $100x^2 - 49 = 0$

5. $81y^2 = 9$

6. $y^2 = 21 - 4y$

7. $y^2 - 7y + 8 = 16$

8. $6x^2 + x - 2 = 0$

9. $3y^2 + y - 2 = 0$

10. $b^2 + 2b - 8 = 0$

11. $4x^2 + 19x - 5 = 0$

12. $8x^2 = 6x + 2$

13. $2y^2 - 6y - 20 = 0$

14. $-6x^2 + 7x - 2 = 0$

15. $y^2 + 3y - 18 = 0$

16. $x^2 + 10x - 11 = 0$

17. $y^2 - 14y + 40 = 0$

18. $b^2 + 9b + 18 = 0$

19. $y^2 - 12y - 13 = 0$

20. $a^2 - 8a - 48 = 0$

21. $x^2 + 2x - 63 = 0$

Solve each of the following quadratic equations by completing the square.

22. $x^2 + 24x + 44 = 0$

23. $x^2 + 6x + 5 = 0$

24. $x^2 - 11x + 5.25 = 0$

Chapter 10
Graphing and Writing Equations and Inequalities

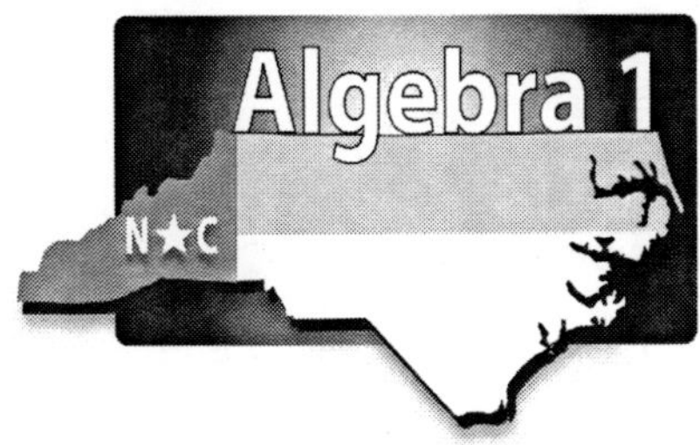

This chapter covers the following North Carolina mathematics standards for Algebra I:

Competency Goal	Objectives
Number and Operations	1.01 1.02
Geometry and Measurement	2.01
Data Analysis and Probability	3.03a 3.03b
Algebra	4.01a 4.01b

10.1 Cartesian Coordinates

A **Cartesian coordinate plane** allows you to graph points with two values. A Cartesian coordinate plane is made up of two number lines. The horizontal number line is called the x-**axis**, and the vertical number line is called the y-**axis**. The point where the x and y axes intersect is called the **origin**. The x and y axes separate the Cartesian coordinate plane into four quadrants that are labeled I, II, III, and IV. The quadrants are labeled and explained on the graph below. Each point graphed on the plane is designated by an **ordered pair** of coordinates. For example, $(2, -1)$ is an ordered pair of coordinates designated by point B on the plane below. The first number, 2, tells you to go over two spaces in a positive direction on the x-axis. The -1 tells you to then go one space down in a negative direction on the y-axis.

Remember: The first number always tells you how far to go right or left of 0, and the second number always tells you how far to go up or down from 0.

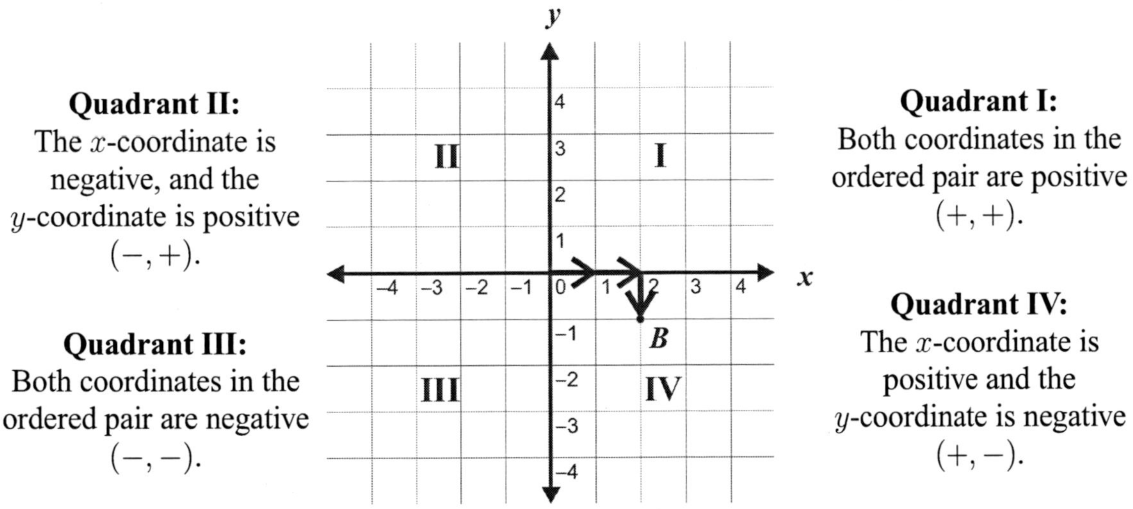

Plot and label the following points on the Cartesian coordinate plane provided.

A. $(2, 4)$	F. $(-3, -5)$	K. $(-1, -1)$	P. $(0, 4)$
B. $(-1, 5)$	G. $(-2, 5)$	L. $(3, -3)$	Q. $(2, 0)$
C. $(3, -4)$	H. $(5, -1)$	M. $(5, 5)$	R. $(-4, 0)$
D. $(-5, -2)$	I. $(4, -4)$	N. $(-2, -2)$	S. $(0, -2)$
E. $(5, 3)$	J. $(5, 2)$	O. $(0, 0)$	T. $(5, 1)$

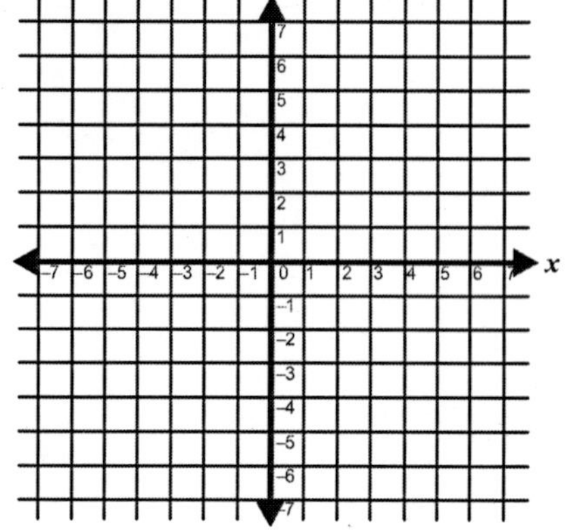

10.2 Identifying Ordered Pairs

When identifying ordered pairs, count how far left or right of 0 to find the x-coordinate and then how far up or down from 0 to find the y-coordinate.

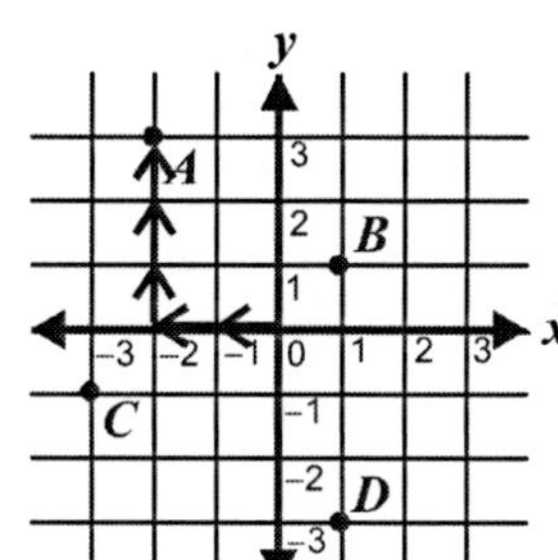

Point A: Left (negative) two and up (positive) three $= (-2, 3)$ in quadrant II

Point B: Right (positive) one and up (positive) one $= (1, 1)$ in quadrant I

Point C: Left (negative) three and down (negative) one $= (-3, -1)$ in quadrant III

Point D: Right (positive one and down (negative) three $= (1, -3)$ in quadrant IV

Find in the ordered pair for each point, and tell which quadrant it is in.

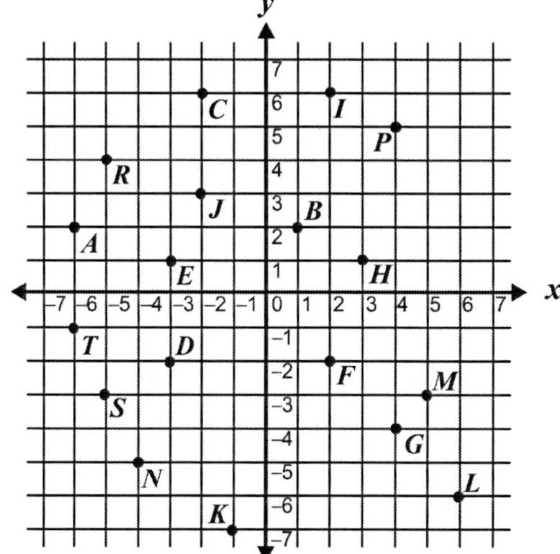

1. point A	4. point D	7. point G	10. point J	13. point M	16. point R
2. point B	5. point E	8. point H	11. point K	14. point N	17. point S
3. point C	6. point F	9. point I	12. point L	15. point P	18. point T

Sometimes, points on a coordinate plane fall on the x or y axis. If a point falls on the x-axis, then the second number of the ordered pair is 0. If a point falls on the y-axis, the first number of the ordered pair is 0.

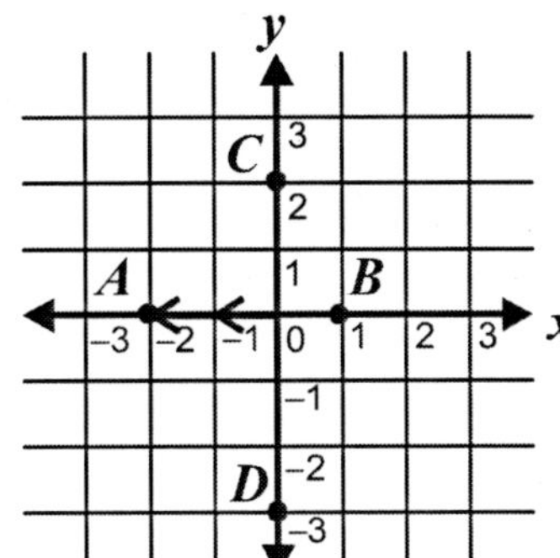

Point A: Left (negative) two and up zero $= (-2, 0)$

Point B: Right (positive) one and up zero $= (1, 0)$

Point C: Left/right zero and up (positive) two $= (0, 2)$

Point D: Left/right zero and down (negative) three $= (0, -3)$

Fill in the ordered pair for each point.

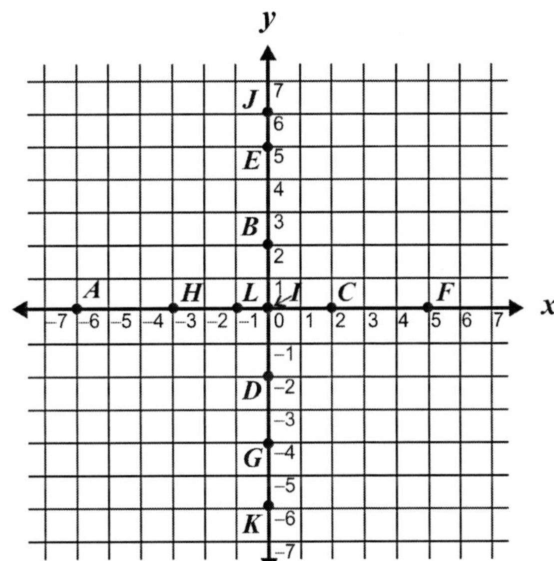

1. point A = (,)

2. point B = (,)

3. point C = (,)

4. point D = (,)

5. point E = (,)

6. point F = (,)

7. point G = (,)

8. point H = (,)

9. point I = (,)

10. point J = (,)

11. point K = (,)

12. point L = (,)

10.3 Finding the Distance Between Two Points

Notice that a subscript added to the x and y identifies each ordered pair uniquely in the plane. For example, point 1 is identified as (x_1, y_1), point 2 as (x_2, y_2), and so on. This unique subscript identification allows us to calculate slope, distance, and midpoints of line segments in the plane using standard formulas like the distance formula. To find the distance between two points on a Cartesian plane, use the following formula:

$$d = \sqrt{(y_2 - y_1)^2 + (x_2 - x_1)^2}$$

Example 1: Find the distance between $(-2, 1)$ and $(3, -4)$.

Step 1: Let $(3, -4) = (x_2, y_2)$ and $(-2, 1) = (x_1, y_1)$.

Plugging the values from the ordered pairs into the formula, we find:

$d = \sqrt{(-4 - 1)^2 + [3 - (-2)]^2}$

$d = \sqrt{(-5)^2 + (5)^2}$

$d = \sqrt{25 + 25}$

$d = \sqrt{50}$

Step 2: To simplify, we look for perfect squares that are a factor of 50. $50 = 25 \times 2$. Therefore,

$d = \sqrt{25} \times \sqrt{2} = 5\sqrt{2}$

Find the distance between the following pairs of points using the distance formula above.

1. $(6, -1)$ $(5, 2)$
2. $(-4, 3)$ $(2, -1)$
3. $(10, 2)$ $(6, -1)$
4. $(-2, 5)$ $(-4, 3)$
5. $(8, -2)$ $(3, -9)$
6. $(2, -2)$ $(8, 1)$
7. $(3, 1)$ $(5, 5)$
8. $(-2, -1)$ $(3, 4)$
9. $(5, -3)$ $(-1, -5)$
10. $(6, 5)$ $(3, -4)$
11. $(-1, 0)$ $(-9, -8)$
12. $(-2, 0)$ $(-6, 6)$
13. $(2, 4)$ $(8, 10)$
14. $(-10, -5)$ $(2, -7)$
15. $(-3, 6)$ $(1, -1)$

10.4 Finding the Midpoint of a Line Segment

You can use the coordinates of the endpoints of a line segment to find the coordinates of the midpoint of the line segment. The formula to find the midpoint between two coordinates is:

$$\text{midpoint, } M = \left(\frac{x_1 + x_2}{2}, \frac{y_1 + y_2}{2}\right)$$

Example 2: Find the midpoint of the line segment having endpoints at $(-3, -1)$ and $(4, 3)$.

Use the formula for the midpoint. $M = \left(\frac{4 + (-3)}{2}, \frac{3 + (-1)}{2}\right)$

When we simplify each coordinate, we find the midpoint, M, is $\left(\frac{1}{2}, 1\right)$.

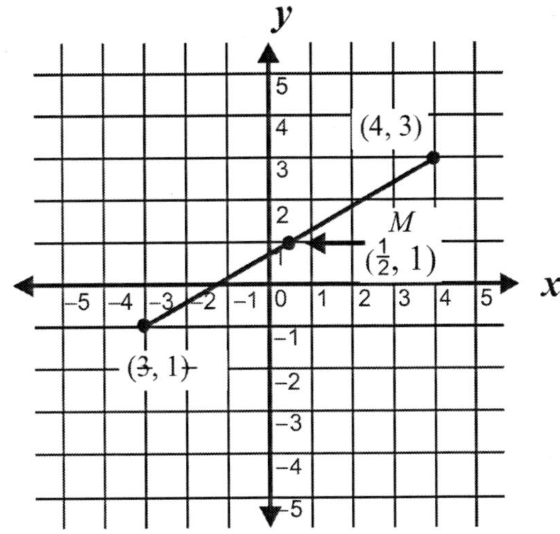

For each of the following pairs of points, find the coordinate of the midpoint, M, using the formula given above.

1. $(4, 5)\ (-6, 9)$
2. $(-3, 2)\ (-1, -2)$
3. $(3, 6)\ (9, 12)$
4. $(2, 5)\ (6, 9)$
5. $(8, 9)\ (6, 11)$
6. $(-4, 3)\ (8, 7)$
7. $(-1, -5)\ (-3, -11)$
8. $(4, 2)\ (-2, 8)$
9. $(4, 3)\ (-1, -5)$
10. $(-6, 2)\ (8, -8)$
11. $(-3, 9)\ (-9, 3)$
12. $(7, 8)\ (11, 6)$
13. $(12, 19)\ (2, 3)$
14. $(5, 4)\ (9, -2)$
15. $(-4, 6)\ (10, -2)$

10.5 Graphing Linear Equations

In addition to graphing ordered pairs, use the Cartesian plane to graph the solution set for an equation. Any equation with two variables that are both to the first power is called a **linear equation.** The graph of a linear equation will always be a straight line.

Example 3: Graph the solution set for $x + y = 7$.

Step 1: Make a list of some pairs of numbers that will work in the equation.

$x + y = 7$		
$4 + 3 = 7$	$(4, 3)$	ordered pair solutions
$-1 + 8 = 7$	$(-1, 8)$	
$5 + 2 = 7$	$(5, 2)$	
$0 + 7 = 7$	$0, 7$	

Step 2: Plot these points on a Cartesian plane.

Step 3: By passing a line through these points, we graph the solution set for $x + y = 7$. This means that every point on the line is a solution to the equation $x + y = 7$. For example, the line passes through the point $(1, 6)$. The point $(1, 6)$ is a solution for the equation. The solutions do not need to be integers. The point $\left(\frac{1}{2}, 6\frac{1}{2}\right)$ is also a solution. Just as there are an infinite number of points on a line, there are an infinite number of solutions to the equation of a line.

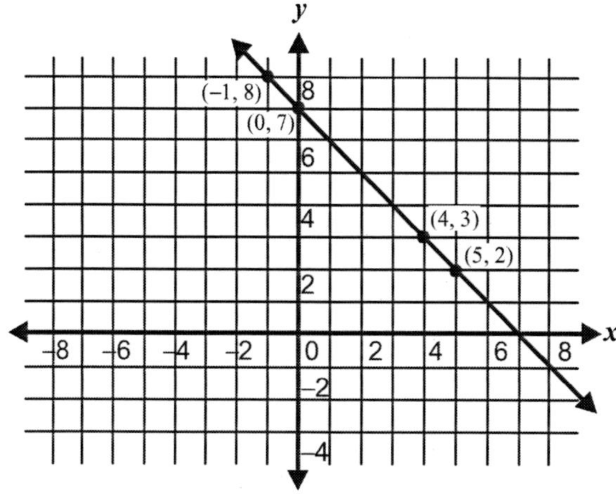

Make a table of solutions for each linear equation below. Then plot the ordered pair solutions on graph paper. Draw a line through the points. (If one of the points does not line up, you have made a mistake.)

1. $x + y = 6$
2. $y = x + 1$
3. $y = x - 2$
4. $x + 2 = y$
5. $x - 5 = y$
6. $x - y = 0$

Example 4: Graph the equation $y = 2x - 5$.

Step 1: This equation has 2 variables, both to the first power, so we know the graph will be a straight line. Substitute some numbers for x or y to find pairs of numbers that satisfy the equation. For the above equation, it will be easier to substitute values of x in order to find the corresponding value for y. Record the values for x and y in a table.

	x	y
If x is 0, y would be -5	0	-5
If x is 1, y would be -3	1	-3
If x is 2, y would be -1	2	-1
If x is 3, y would be 1	3	1

Step 2: Graph the ordered pairs, and draw a line through the points.

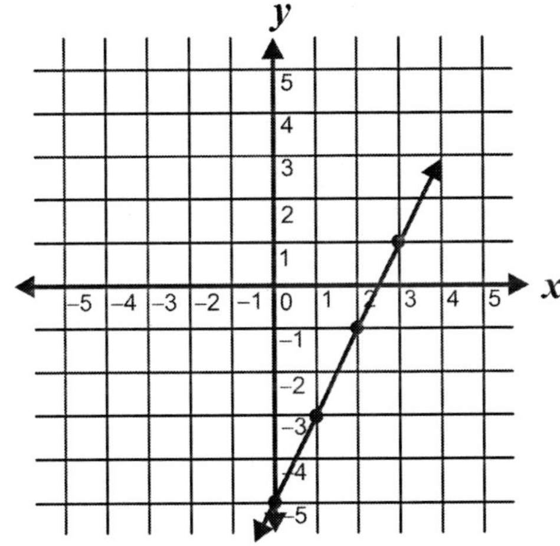

Find pairs of numbers that satisfy the equations below, and graph the line on graph paper.

1. $y = -2x + 2$
2. $2x - 2 = y$
3. $-x + 3 = y$
4. $y = x + 1$
5. $4x - 2 = y$
6. $y = 3x - 3$
7. $x = 4y - 3$
8. $2x = 3y + 1$
9. $x + 2y = 4$

10.6 Graphing Horizontal and Vertical Lines

The graph of some equations is a horizontal or a vertical line.

Example 5: $y = 3$

Step 1: Make a list of ordered pairs that satisfy the equation $y = 3$.

x	y
0	3
1	3
2	3
3	3

No matter what value of x you choose, y is always 3.

Step 2: Plot these points on an Cartesian plane, and draw a line through the points.

The graph is a horizontal line.

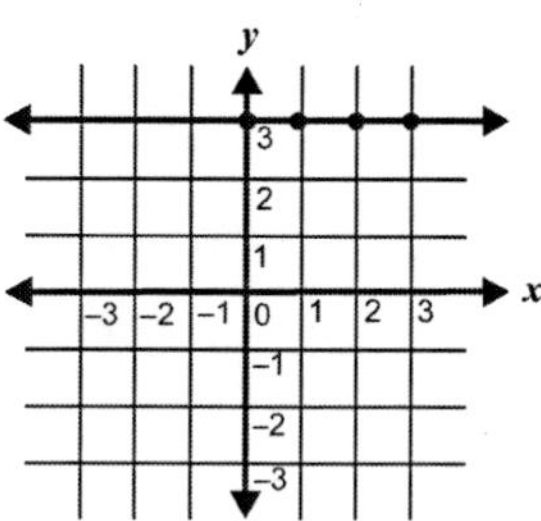

Example 6: $2x + 3 = 0$

Step 1: For these equations with only one variable, find what x equals first.

$2x + 3 = 0$
$2x = -3$
$x = -\frac{3}{2}$

Just like Example 3, find ordered pairs that satisfy the equation, plot the points, and graph the line.

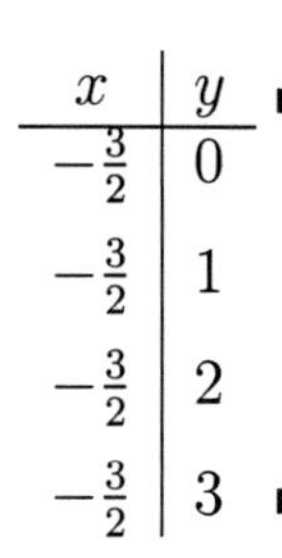

x	y
$-\frac{3}{2}$	0
$-\frac{3}{2}$	1
$-\frac{3}{2}$	2
$-\frac{3}{2}$	3

No matter which value of y you choose, the value of x does not change.

The graph is a vertical line.

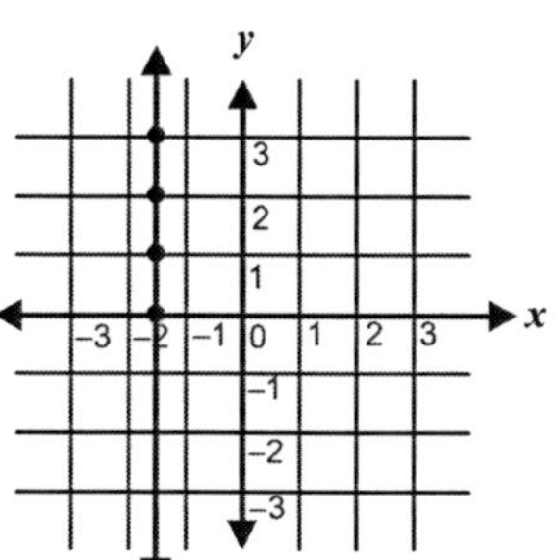

Find pairs of numbers that satisfy the equations below, and graph the line on graph paper.

1. $2y + 2 = 0$
2. $x = -4$
3. $3x = 3$
4. $y = 5$
5. $4x - 2 = 0$
6. $2x - 6 = 0$
7. $4y = 1$
8. $5x + 10 = 0$
9. $3y + 12 = 0$
10. $x + 1 = 0$
11. $2y - 8 = 0$
12. $3x = -9$
13. $x = -2$
14. $6y - 2 = 0$
15. $5x - 5 = 0$

10.7 Finding the Intercepts of a Line

The x-intercept is the point where the graph of a line crosses the x-axis. The y-intercept is the point where the graph of a line crosses the y-axis.

To find the x-intercept, set $y = 0$
To find the y-intercept, set $x = 0$

Example 7: Find the x- and y-intercepts of the line $6x + 2y = 18$

Step 1: To find the x-intercept, set $y = 0$.

$$\begin{aligned} 6x + 2(0) &= 18 \\ \frac{6x}{6} &= \frac{18}{6} \\ x &= 3 \end{aligned}$$

The x-intercept is at the point $(3, 0)$.

Step 2: To find the y-intercept, set $x = 0$.

$$\begin{aligned} 6(0) + 2y &= 18 \\ \frac{2y}{2} &= \frac{18}{2} \\ y &= 9 \end{aligned}$$

The y-intercept is at the point $(0, 9)$.

Step 3: You can now use the two intercepts to graph the line.

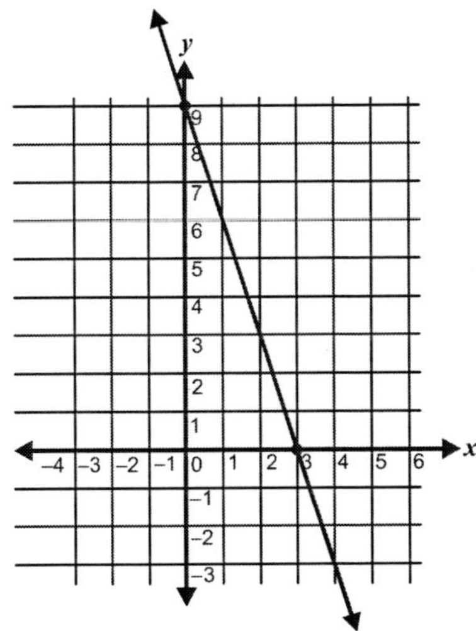

10.8 Understanding Slope

The slope of a line refers to how steep a line is. Slope is also defined as the rate of change. When we graph a line using ordered pairs, we can easily determine the slope. Slope is often represented by the letter m.

> The formula for slope of a line is: $m = \dfrac{y_2 - y_1}{x_2 - x_1}$ or $\dfrac{\text{rise}}{\text{run}}$

Example 8: What is the slope of the following line that passes through the ordered pairs $(-4, -3)$ and $(1, 3)$?

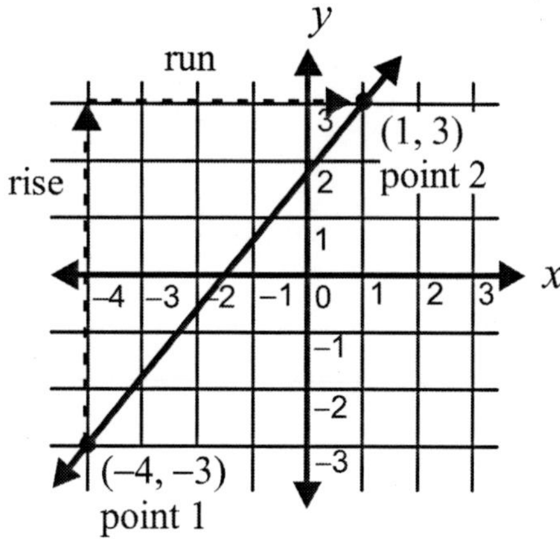

y_2 is 3, the y-coordinate of point 2.

y_1 is -3, the y-coordinate of point 1.

x_2 is 1, the x-coordinate of point 2.

x_1 is -4, the x-coordinate of point 1.

Use the formula for slope given above: $m = \dfrac{3-(-3)}{1-(-4)} = \dfrac{6}{5}$

The slope is $\frac{6}{5}$. This shows us that we can go up 6 (rise) and over 5 to the right (run) to find another point on the line.

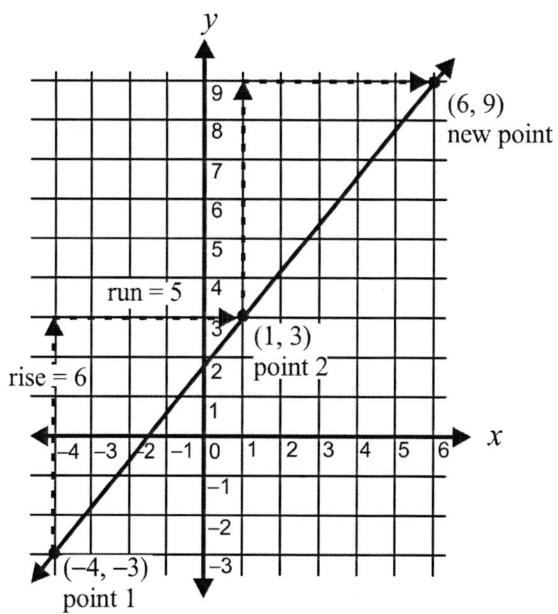

Example 9: Find the slope of a line through the points $(-2, 3)$ and $(1, -2)$. It doesn't matter which pair we choose for point 1 and point 2. The answer is the same.

Let point 1 be $(-2, 3)$
Let point 2 be $(1, -2)$

$$\text{slope} = \frac{(y_2 - y_1)}{(x_2 - x_1)} = \frac{-2 - 3}{1 - (-2)} = \frac{-5}{3}$$

When the slope is negative, the line will slant up and to the left. For this example, the line will go **down** 5 units and then over 3 units to the **right**.

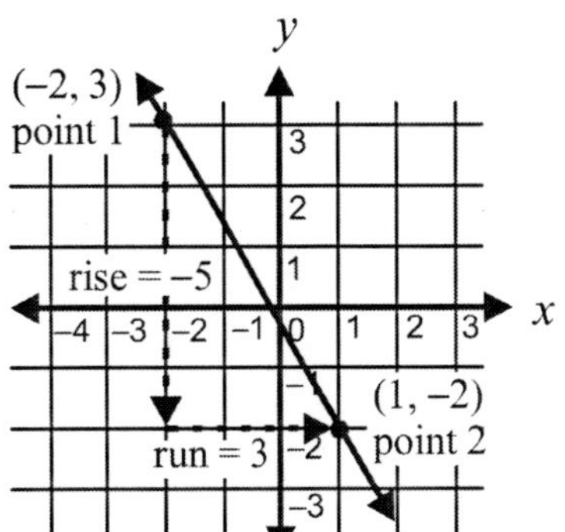

Example 10: What is the slope of a line that passes through $(1, 1)$ and $(3, 1)$?

$$\text{slope} = \frac{1 - 1}{3 - 1} = \frac{0}{2} = 0$$

When $y_2 - y_1 = 0$, the slope will equal 0, and the line will be horizontal.

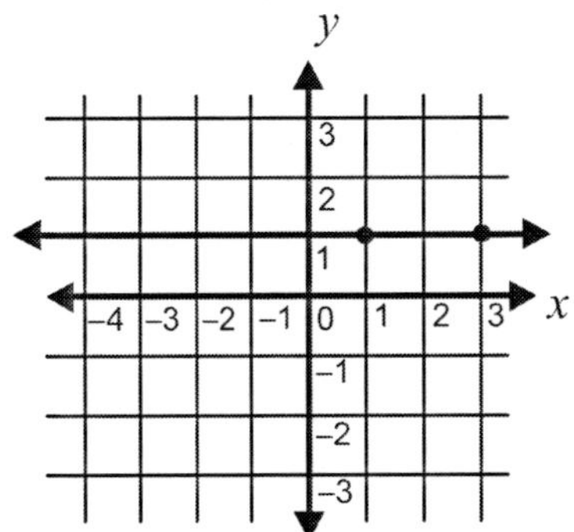

Example 11: What is the slope of a line that passes through $(2, 1)$ and $(2, -3)$?

$$\text{slope} = \frac{-3 - 1}{2 - 2} = \frac{4}{0} = \text{undefined}$$

When $x_2 - x_1 = 0$, the slope is undefined, and the line will be vertical.

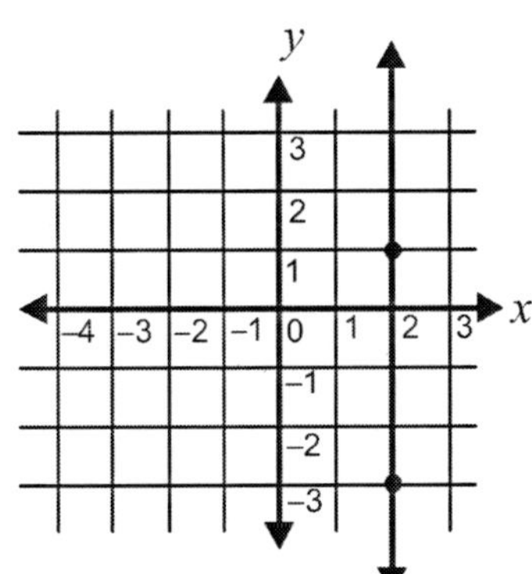

The following lines summarize what we know about slope.

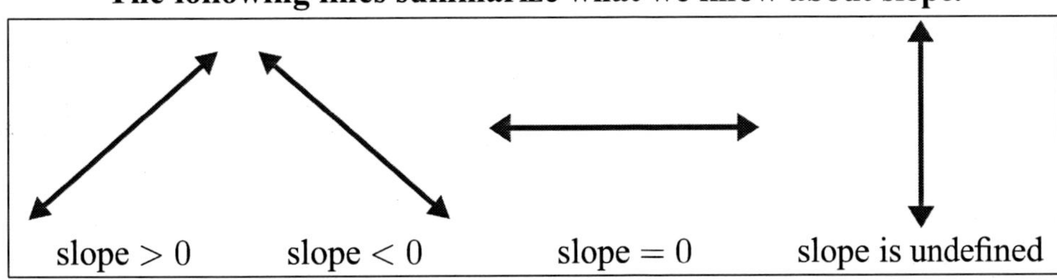

Find the slope of the line that goes through the following pairs of points. Then, using graph paper, graph the line through the two points, and label the rise and run.
(See Examples 8 – 11)

Use the formula slope $= \dfrac{y_2 - y_1}{x_2 - x_1}$**.**

1. $(2, 3)$ $(4, 5)$
2. $(1, 3)$ $(2, 5)$
3. $(-1, 2)$ $(4, 1)$
4. $(1, -2)$ $(4, -2)$
5. $(3, 0)$ $(3, 4)$
6. $(3, 2)$ $(-1, 8)$
7. $(4, 3)$ $(2, 4)$
8. $(2, 2)$ $(1, 5)$
9. $(3, 4)$ $(1, 2)$
10. $(3, 2)$ $(3, 6)$
11. $(6, -2)$ $(3, -2)$
12. $(1, 2)$ $(3, 4)$
13. $(-2, 1)$ $(-4, 3)$
14. $(5, 2)$ $(4, -1)$
15. $(1, -3)$ $(-2, 4)$
16. $(2, -1)$ $(3, 5)$

10.9 Slope-Intercept Form of a Line

An equation that contains two variables, each to the first degree, is a **linear equation**. The graph for a linear equation is a straight line. To put a linear equation in slope-intercept form, solve the equation for y. This form of the equation shows the slope and the y-intercept. Slope-intercept form follows the pattern of $y = mx + b$. The "m" represents slope, and the "b" represents the y-intercept. The y-intercept is the point at which the line crosses the y-axis. The constant is called the **slope** of the line.

Example 12: Put the equation $2x + 3y = 15$ in slope-intercept form. What is the slope of the line? What is the y-intercept? Graph the line.

Step 1: Solve for y:

$$\begin{array}{rcrcl} 2x & + & 3y & = & 15 \\ -2x & & & & -2x \\ \hline & & \frac{3y}{3} & = & -\frac{2x}{3} + \frac{15}{3} \end{array}$$

slope-intercept form: $y = -\frac{2}{3}x + 5$

The slope is $-\frac{2}{3}$ and the y-intercept is 5.

Step 2: Knowing the slope and the y-intercept, we can graph the line.

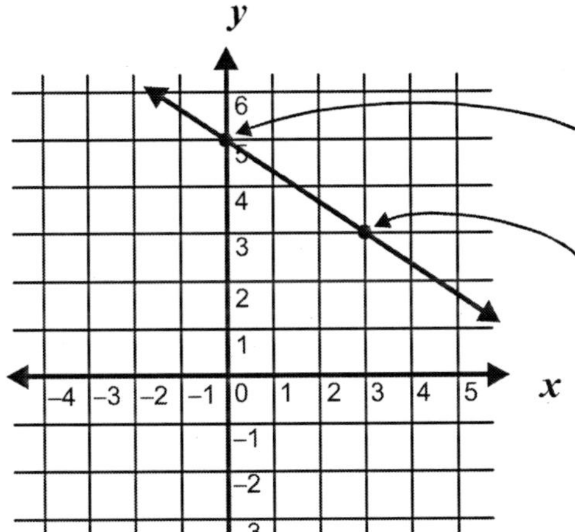

The y-intercept is 5, so the line passes through the point $(0, 5)$ on the y-axis.

The slope is $-\frac{2}{3}$, so go down 2 and over 3 to get a second point.

Put each of the following equations in slope-intercept form by solving for y. On your graph paper, graph the line using the slope and y-intercept.

1. $4x - 5y = 5$
2. $2x + 4y = 16$
3. $3x - 2y = 10$
4. $x + 3y = -12$
5. $6x + 2y = 0$
6. $8x - 5y = 10$
7. $-2x + y = 4$
8. $-4x + 3y = 12$
9. $-6x + 2y = 12$
10. $x - 5y = 5$
11. $3x - 2y = -6$
12. $3x + 4y = 2$
13. $-x = 2 + 4y$
14. $2x = 4y - 2$
15. $6x - 3y = 9$
16. $4x + 2y = 8$
17. $6x - y = 4$
18. $-2x - 4y = 8$
19. $5x + 4y = 16$
20. $6 = 2y - 3x$

10.10 Verify That a Point Lies on a Line

To know whether or not a point lies on a line, substitute the coordinates of the point into the equation for the line. If the point lies on the line, the equation will be true. If the point does not lie on the line, the equation will be false.

Example 13: Does the point $(5, 2)$ lie on the line given by the equation $x + y = 7$?

Solution: Substitute 5 for x and 2 for y in the equation. $5 + 2 = 7$. Since this is a true statement, the point $(5, 2)$ does lie on the line $x + y = 7$.

Example 14: Does the point $(0, 1)$ lie on the line given by the equation $5x + 4y = 16$?

Solution: Substitute 0 for x and 1 for y in the equation $5x + 4y = 16$. Does $5\,(0) + 4\,(1) = 16$? No, it equals 4, not 16. Therefore, the point $(0, 1)$ is not on the line given by the equation $5x + 4y = 16$.

For each point below, state whether or not it lies on the line given by the equation that follows the point coordinates.

1. $(2, 4)$ $6x - y = 8$
2. $(1, 1)$ $6x - y = 5$
3. $(3, 8)$ $-2x + y = 2$
4. $(9, 6)$ $-2x + y = 0$
5. $(3, 7)$ $x - 5y = -32$
6. $(0, 5)$ $-6x - 5y = 3$
7. $(2, 4)$ $4x + 2y = 16$
8. $(9, 1)$ $3x - 2y = 29$
9. $(6, 8)$ $6x - y = 28$
10. $(-2, 3)$ $x + 2y = 4$
11. $(4, -1)$ $-x - 3y = -1$
12. $(-1, -3)$ $2x + y = 1$

10.11 Graphing a Line Knowing a Point and Slope

If you are given a point of a line and the slope of a line, you can graph the line.

Example 15: Given that line l has a slope of $\frac{4}{3}$ and contains the point $(2, -1)$, graph the line.

Plot and label the point $(2, -1)$ on a Cartesian plane.

The slope, m, is $\frac{4}{3}$, so the rise is 4, and the run is 3. From the point $(2, -1)$, count 4 units up and 3 units to the right.

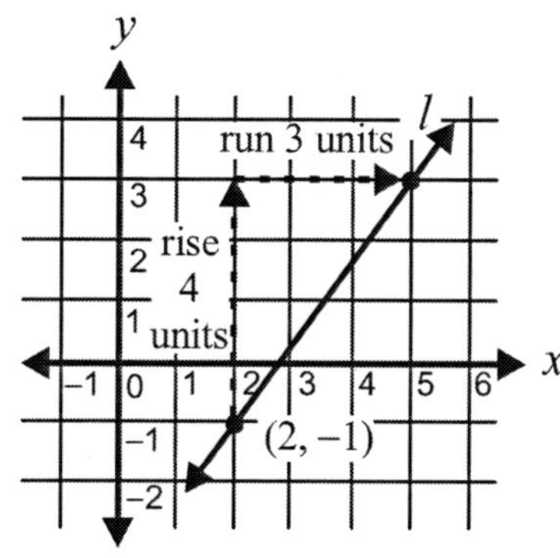

Draw the line through the two points.

Example 16: Given a line that has a slope of $-\frac{1}{4}$ and passes through the point $(-3, 2)$, graph the line.

Plot the point $(-3, 2)$.

Since the slope is negative, go **down** 1 unit and over 4 units to get a second point.

Graph the line through the two points.

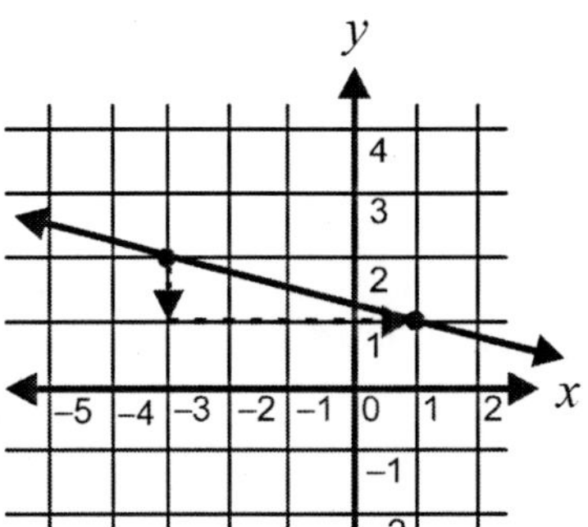

Graph a line on your own graph paper for each of the following problems. First, plot the point. Then use the slope to find a second point. Draw the line formed from the point and the slope.

1. $(2, -2), m = \frac{3}{4}$
2. $(3, -4), m = \frac{1}{2}$
3. $(1, 3), m = -\frac{1}{3}$
4. $(2, -4), m = 1$
5. $(3, 0), m = -\frac{1}{2}$
6. $(-2, 1), m = \frac{4}{3}$
7. $(-4, -2), m = \frac{1}{2}$
8. $(1, -4), m = \frac{3}{4}$
9. $(2, -1), m = -\frac{1}{2}$
10. $(5, -2), m = \frac{1}{4}$
11. $(-2, -3), m = \frac{2}{3}$
12. $(4, -1), m = -\frac{1}{3}$
13. $(-1, 5), m = \frac{2}{5}$
14. $(-2, 3), m = \frac{3}{4}$
15. $(4, 4), m = -\frac{1}{2}$
16. $(3, -3), m = -\frac{3}{4}$
17. $(-2, 5), m = \frac{1}{3}$
18. $(-2, -3), m = -\frac{3}{4}$
19. $(4, -3), m = \frac{2}{3}$
20. $(1, 4), m = -\frac{1}{2}$

10.12 Finding the Equation of a Line Using Two Points or a Point and Slope

If you know the slope of a line, and you know the coordinates of one point, you can write the equation for the line. You know the formula for the slope of a line is:

$$m = \frac{y_2 - y_1}{x_2 - x_1} \text{ or } \frac{y_2 - y_1}{x_2 - x_1} = m$$

Using algebra, you can see that if you multiply both sides of the equation by $x_2 - x_1$, you get:

$$y - y_1 = m(x - x_1) \longleftarrow \text{point-slope form of an equation}$$

Example 17: Write the equation of the line passing through the points $(-2, 3)$ and $(1, 5)$.

Step 1: First, find the slope of the line using the two points given.

$$m = \frac{y_2 - y_1}{x_2 - x_1} = \frac{5 - 3}{1 - (-2)} = \frac{2}{3}$$

Step 2: Pick one of the two points to use in the point-slope equation. For point $(-2, 3)$, we know $x_1 = -2$ and $y_1 = 3$, and we know $m = \frac{2}{3}$. Substitute these values into the point-slope form of the equation.

$$y - y_1 = m(x - x_1)$$
$$y - 3 = \frac{2}{3}[x - (-2)]$$
$$y - 3 = \tfrac{2}{3}x + \tfrac{4}{3}$$
$$y = \tfrac{2}{3}x + \tfrac{13}{3}$$

Use the point-slope formula to write an equation for each of the following lines.

1. $(1, -2)$, $m = 2$
2. $(-3, 3)$, $m = \frac{1}{3}$
3. $(4, 2)$, $m = \frac{1}{4}$
4. $(5, 0)$, $m = 1$
5. $(3, -4)$, $m = \frac{1}{2}$
6. $(-1, -4)$ $(2, -1)$
7. $(2, 1)$ $(-1, -3)$
8. $(-2, 5)$ $(-4, 3)$
9. $(-4, 3)$ $(2, -1)$
10. $(3, 1)$ $(5, 5)$
11. $(-3, 1)$, $m = 2$
12. $(-1, 2)$, $m = \frac{4}{3}$
13. $(2, -5)$, $m = -2$
14. $(-1, 3)$, $m = \frac{1}{3}$
15. $(0, -2)$, $m = -\frac{3}{2}$

10.13 Graphing Inequalities

In a previous section, you would graph the equation $x = 3$ as:

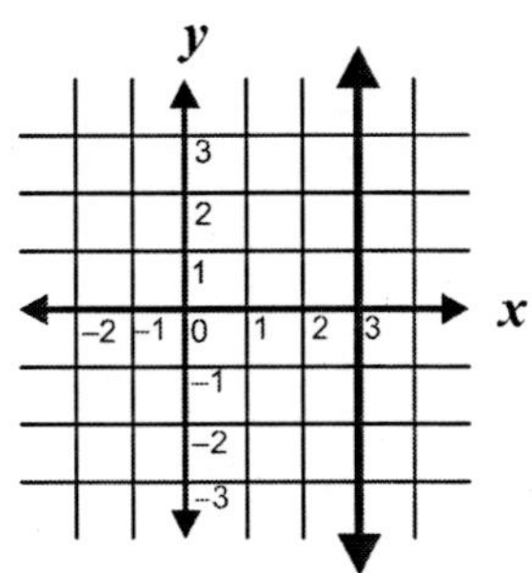

In this section, we graph inequalities such as $x > 3$ (read x is greater than 3). To show this, we use a broken line since the points on the line $x = 3$ are not included in the solution. We shade all points greater than 3.

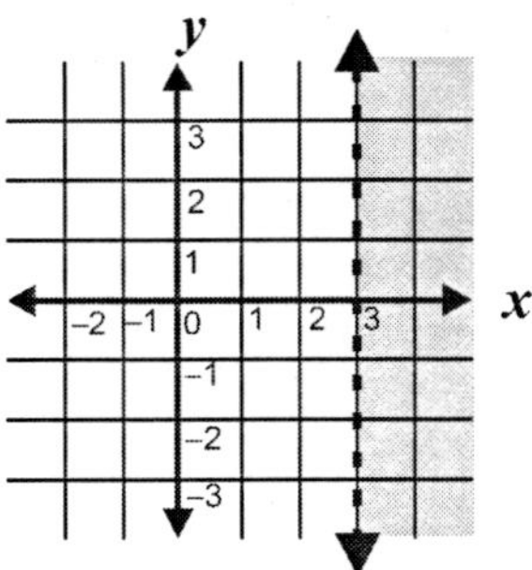

When we graph $x \geq 3$ (read x is greater than or equal to 3), we use a solid line because the points on the line $x = 3$ are included in the graph.

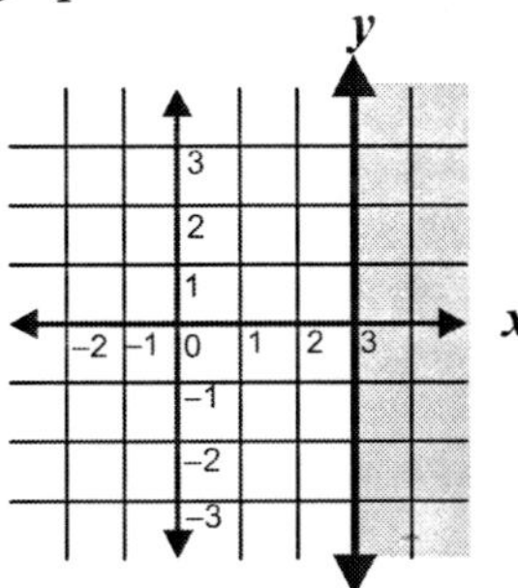

Graph the following inequalities on your own graph paper.

1. $y < 2$
2. $x \geq 4$
3. $y \geq 1$
4. $x < -1$
5. $y \geq -2$
6. $x \leq -4$
7. $x > -3$
8. $y \leq 3$
9. $x \leq 5$
10. $y > -5$
11. $x \geq 3$
12. $y < -1$
13. $x \leq 0$
14. $y > -1$
15. $y \leq 4$
16. $x \geq 0$
17. $y \geq 3$
18. $x < 4$
19. $x \leq -2$
20. $y < -2$
21. $y \geq -4$
22. $x \geq -1$
23. $y \leq 5$
24. $x < -3$

Example 18: Graph $x + y \geq 3$.

Step 1: First, we graph $x + y \geq 3$ by changing the inequality to an equality. Think of ordered pairs that will satisfy the equation $x + y = 3$. Then, plot the points, and draw the line. As shown below, this line divides the Cartesian plane into 2 half-planes, $x + y \geq 3$ and $x + y \leq 3$. One half-plane is above the line, and the other is below the line.

x	y
2	1
0	3
3	0
4	−1

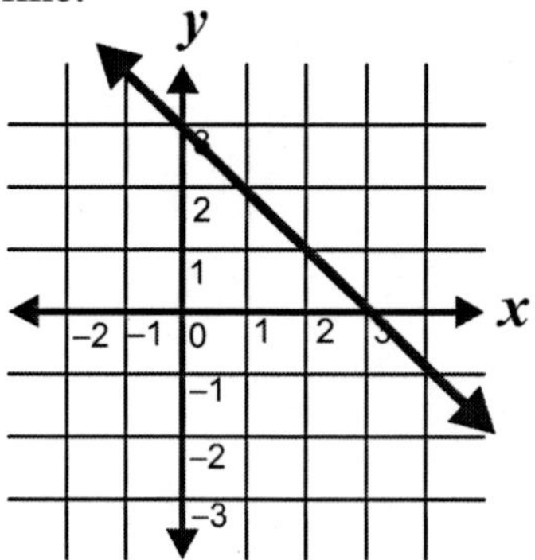

Step 2: To determine which side of the line to shade, first choose a test point. If the point you choose makes the inequality true, then the point is on the side you shade. If the point you choose does not make the inequality true, then shade the side that does not contain the test point.

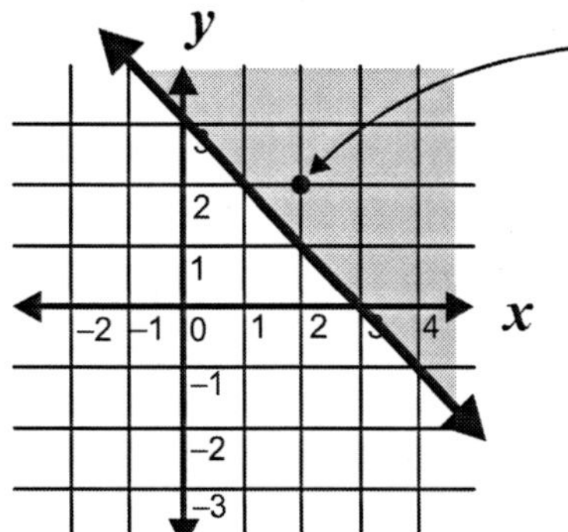

For our test point, let's choose $(2, 2)$. Substitute $(2, 2)$ into the inequality.

$x + y \geq 3$
$2 + 2 \geq 3$

$4 \geq 3$ is true, so shade the side that includes this point.

Use a solid line because of the $\geq$ sign.

Graph the following inequalities on your own graph paper.

1. $x + y \leq 4$
2. $x + y \geq 3$
3. $x \geq 5 - y$
4. $x \leq 1 + y$
5. $x - y \geq -2$
6. $x < y + 4$
7. $x + y < -1$
8. $x - y \leq 0$
9. $x \geq y + 2$
10. $x < -y + 1$
11. $-x + y > 1$
12. $-x - y < -2$

For more complex inequalities, it is easier to graph by first changing the inequality to an equality and then put the equation in slope-intercept form.

Example 19: Graph the inequality $2x + 4y \leq 8$.

Step 1: Change the inequality symbol to an equal sign.
$2x + 4y = 8$

Step 2: Put the equation in slope-intercept form by solving the equation for y.

$$2x + 4y = 8$$

$$2x - 2x + 4y = -2x + 8 \quad \text{Subtract } 2x \text{ from both sides of the equation.}$$

$$4y = -2x + 8 \quad \text{Simplify.}$$

$$\frac{4y}{4} = \frac{-2x + 8}{4} \quad \text{Divide both sides by 4.}$$

$$y = \frac{-2x}{4} + \frac{8}{4} \quad \text{Find the lowest terms of the fractions.}$$

$$y = -\tfrac{1}{2}x + 2$$

Step 3: Graph the line. If the inequality is $<$ or $>$, use a dotted line. If the inequality is $\leq$ or $\geq$, use a solid line. For this example, we should use a solid line.

Step 4: Determine which side of the line to shade. Pick a point such as $(0, 0)$ to see if it is true in the inequality.

$2x + 4y \leq 8$, so substitute $(0, 0)$.
Is $0 + 0 \leq 8$? Yes, $0 \leq 8$, so shade the side of the line that includes the point $(0, 0)$.

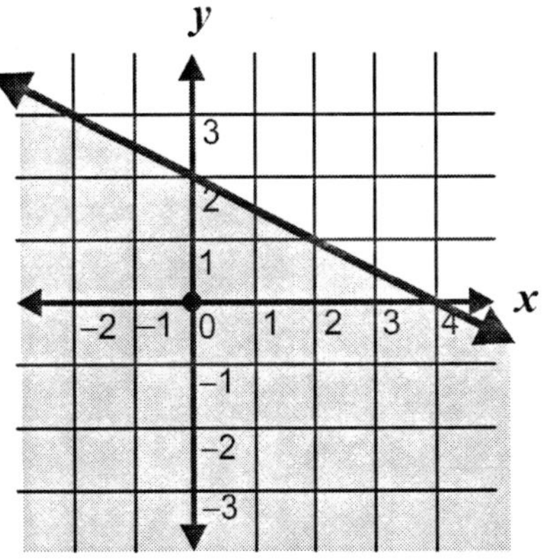

Graph the following inequalities on your own graph paper.

1. $2x + y \geq 1$
2. $3x - y \leq 3$
3. $x + 3y > 12$
4. $4x - 3y < 12$
5. $y \geq 3x + 1$
6. $x - 2y > -2$
7. $x \leq y + 4$
8. $x + y < -1$
9. $-4y \geq 2x + 1$
10. $x \leq 4y - 2$
11. $3x - y \geq 4$
12. $y \geq 2x - 5$
13. $x + 7y < 1$
14. $-2y < 4x - 1$
15. $y > 4x + 1$

Chapter 10 Review

Record the coordinates and quadrants of the following points.

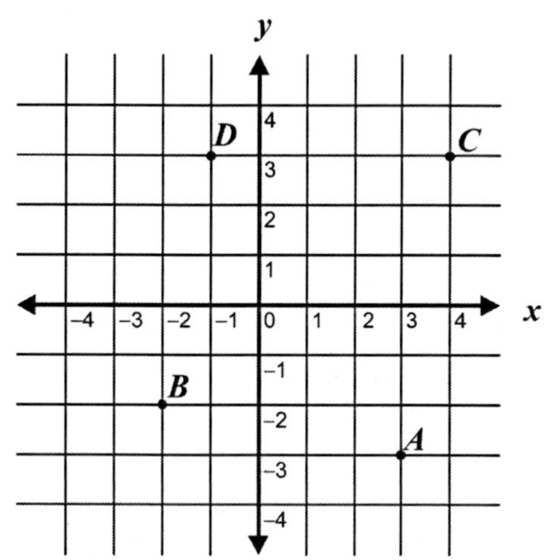

	Coordinates	Quadrants
1. $A =$	_______	_______
2. $B =$	_______	_______
3. $C =$	_______	_______
4. $D =$	_______	_______

On the same plane above, label these additional coordinates.

5. $E = (0, -3)$
6. $F = (-3, -1)$
7. $G = (4, 0)$
8. $H = (2, 2)$

Answer the following questions.

9. In which quadrant does the point $(2, 3)$ lie?

10. In which quadrant does the point $(-5, -2)$ lie?

11. Graph the solution set for the linear equation: $x - 3 = y$ on your own graph paper.

12. Which of the following is not a solution of $3x = 5y - 1$?

(A) $(3, 2)$
(B) $(7, 4)$
(C) $\left(-\frac{1}{3}, 0\right)$
(D) $(-2, -1)$

13. The point $(-2, 1)$ is a solution for which of the following equations?

(A) $y + 2x = 4$
(B) $-2x - y = 5$
(C) $x + 2y = -4$
(D) $2x - y = -5$

14. Graph the equation $2x - 4 = 0$ on your own graph paper.

15. What is the slope of the line that passes through the points $(5, 3)$ and $(6, 1)$?

16. What is the slope of the line that passes through the points $(-1, 4)$ and $(-6, -2)$?

17. What is the x-intercept for the following equation? $6x - y = 30$

18. What is the y-intercept for the following equation? $4x + 2y = 28$

19. Graph the equation $3y = 9$ on your own graph paper.

20. Write the following equation in slope-intercept form.
$$3x = -2y + 4$$

21. What is the slope of the line $y = -\frac{1}{2}x + 3$?

22. What is the x-intercept of the line $y = 5x + 6$?

23. What is the y-intercept of the line $y - \frac{2}{3}x + 3 = 0$?

24. Graph the line which has a slope of -2 and a y-intercept of -3 on your own graph paper.

25. Which of the following points does **not** lie on the line $y = 3x - 2$?

    (A) $(0, -2)$
    (B) $(1, 1)$
    (C) $(-1, 5)$
    (D) $(2, 4)$

26. Find the equation of the line which contains the point $(0, 2)$ and has a slope of $\frac{3}{4}$.

27. Which is the graph of $x - 3y = 6$?

(A)

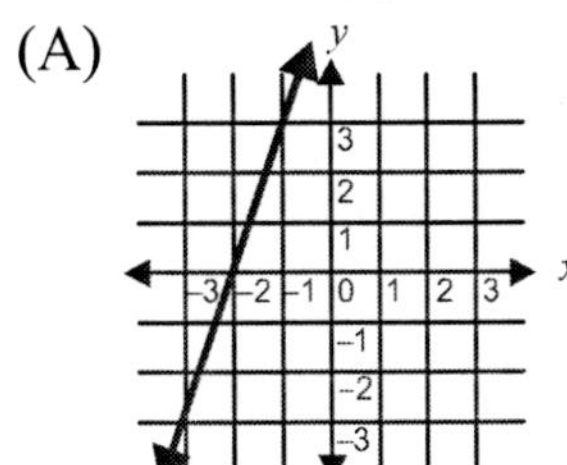

(B)

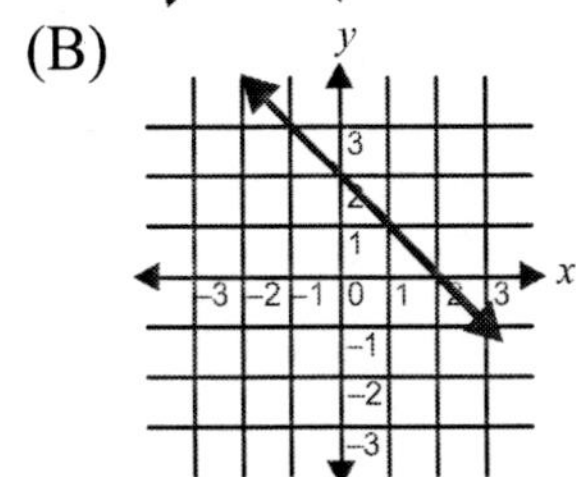

(C)

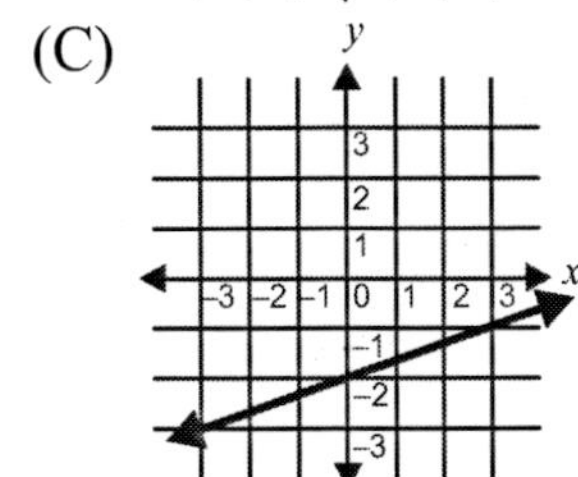

(D)

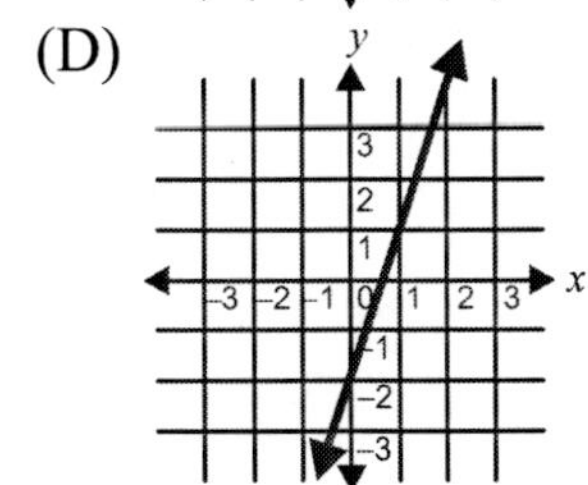

Graph the following inequalities on a Cartesian plane using your graph paper.

28. $x \geq 4$

29. $x \leq -2$

30. $5y > -10x + 5$

31. $y \leq 2$

32. $2x + y \leq 5$

33. $y - 2x \leq 3$

34. $y \geq x + 2$

35. $3 + y > x$

36. What is the distance between the points $(3, 3)$ and $(6, -1)$?

37. What is the distance between the two points $(-3, 0)$ and $(2, 5)$?

For questions 38 and 39, use the following formula to find the coordinates of the midpoint of the line segments with the given endpoints.

$$\text{midpoint} = \left(\frac{x_1 + x_2}{2}, \frac{y_1 + y_2}{2}\right)$$

38. $(6, 10)$ $(-4, 4)$

39. $(-1, -7)$ $(5, 3)$

Chapter 11
Applications of Graphs

This chapter covers the following North Carolina mathematics standards for Algebra I:

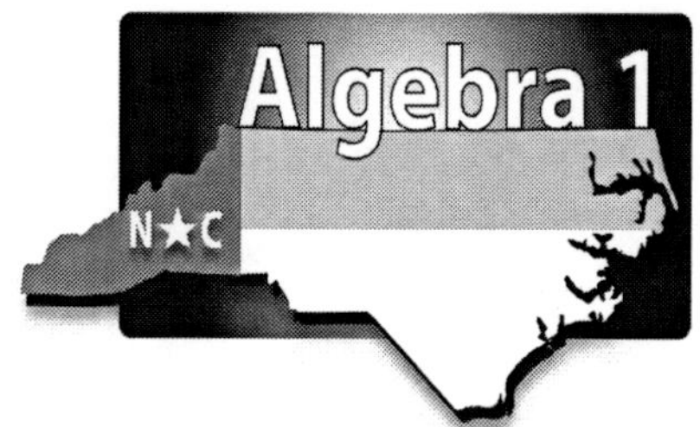

Competency Goal	Objectives
Number and Operations	1.01 1.02
Geometry and Measurement	2.02
Data Analysis and Probability	3.03a, b
Algebra	4.01a 4.02

11.1 Changing the Slope or Y-Intercept of a Line

When the slope and/or the y-intercept of a linear equation changes, the graph of the line will also change.

Example 1: Consider line l shown in Figure 1 at right. What happens to the graph of the line if the slope is changed to $\frac{4}{5}$?

Determine the y-intercept of the line. For line l, it can easily be seen from the graph that the y-intercept is at the point $(0, -1)$.

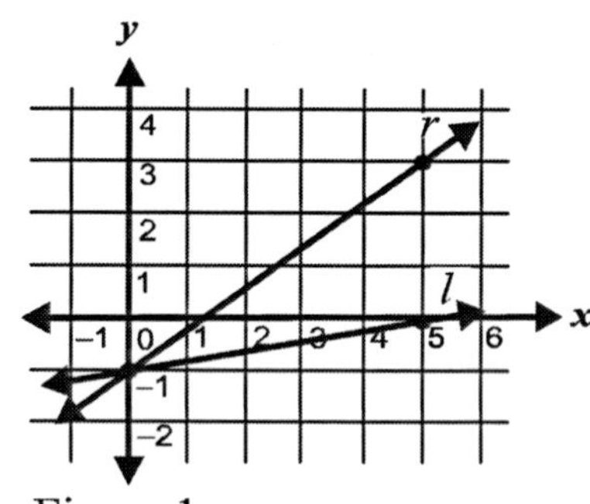

Figure 1

Find the slope of the line using two points that the line goes through: $(0, -1)$ and $(5, 0)$.

$$m = \frac{y_2 - y_1}{x_2 - x_2} = \frac{0 - (-1)}{5 - 0} = \frac{1}{5}$$

Write the equation of line l in slope-intercept form:

$y = mx + b \quad \Longrightarrow \quad y = \frac{1}{5}x - 1$

Rewrite the equation of the line using a slope of $\frac{4}{5}$, and then graph the line. The equation of the new line is $y = \frac{4}{5}x - 1$.

The graph of the new line is labeled line r and is shown in Figure 1. A line with a slope of $\frac{4}{5}$ is steeper than a line with a slope of $\frac{1}{5}$.

Note: The greater the numerator, or "rise," of the slope, the steeper the line will be. The greater the denominator, or "run," of the slope, the flatter the line will be.

Example 2: Consider line l shown in Figure 2 below. The equation of the line is $y = -\frac{1}{2}x + 3$. What happens to the graph of the line if the y-intercept is changed to -1?

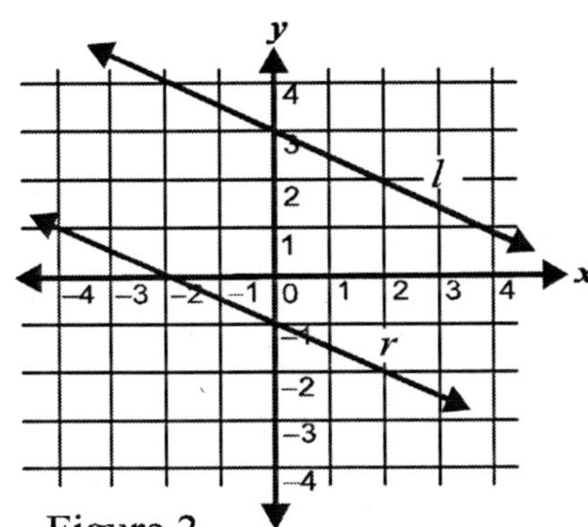

Figure 2

Rewrite the equation of the line replacing the y-intercept with -1. The equation of the new line is $y = -\frac{1}{2}x - 1$.

Graph the new line. Line r in Figure 2 is the graph of the equation $y = -\frac{1}{2}x - 1$. Since both lines l and r have the same slope, they are parallel. Line r, with a y-intercept of -1, sits below line l, with a y-intercept of 3.

Put each pair of the following equations in slope-intercept form. Write P if the lines are parallel and NP if the lines are not parallel.

1. $y = x + 1$ \
   $2y - 2x = 6$ _____

2. $2x + y = 6$ \
   $2x = 8 - y$ _____

3. $x + 5y = 0$ \
   $5y + 5 = x$ _____

4. $y = 3 - \frac{1}{3}x$ \
   $3y + x = -6$ _____

5. $x = 2y$ \
   $-x = -2y + 14$ _____

6. $y = x + 2$ \
   $-y = x + 4$ _____

7. $y = 4 - \frac{1}{4}x$ \
   $3x + 4y = 4$ _____

8. $x + y = 5$ \
   $5 - y = 2x$ _____

9. $x - 4y = 0$ \
   $4y = x - 8$ _____

Consider the line (l) shown on each of the following graphs, and write the equation of the line in the space provided. Then, on the same graph, graph the line (r) for which the equation is given. Write how the slope and y-intercept of line l compare to the slope and y-intercept of line r for each graph.

10.

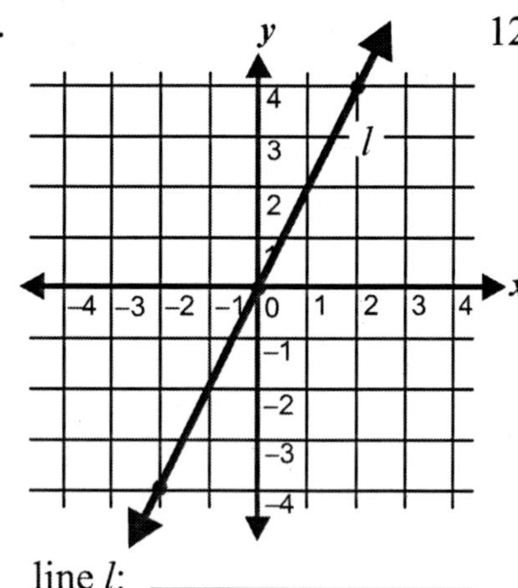

line l: ________________

line r: $y = -2x$

slopes: ________________

y-intercepts: ________________

11.

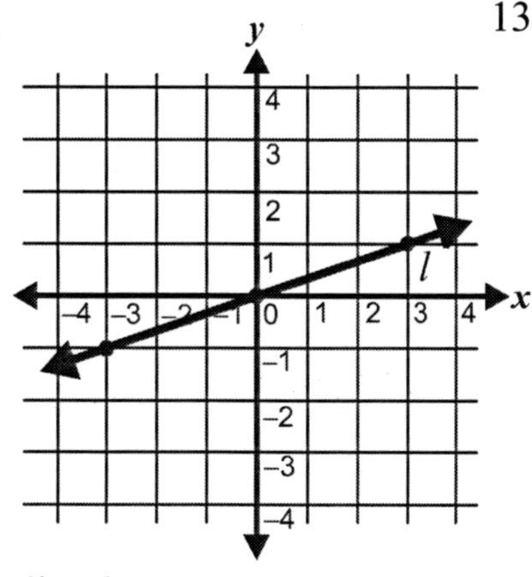

line l: ________________

line r: $y = \frac{1}{3}x + 2$

slopes: ________________

y-intercepts: ________________

12.

y

l

x

-4 -3 -2 -1 0 1 2 3 4

4 3 2 1 -1 -2 -3 -4

line l: ________________

line r: $y = -3x - 1$

slopes: ________________

y-intercepts: ________________

13.

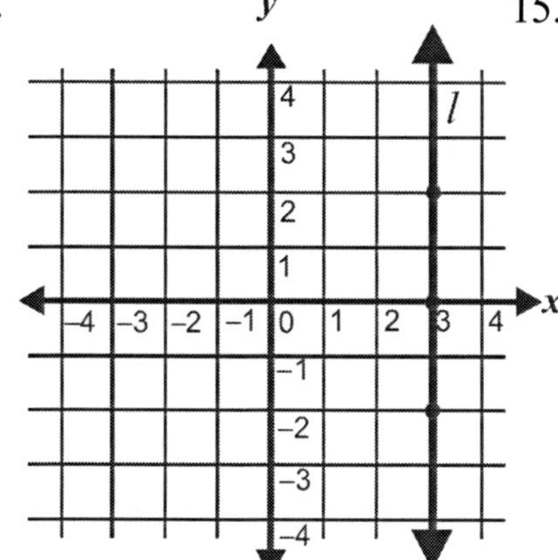

line l: ________________

line r: $y = -3$

slopes: ________________

y-intercepts: ________________

14.

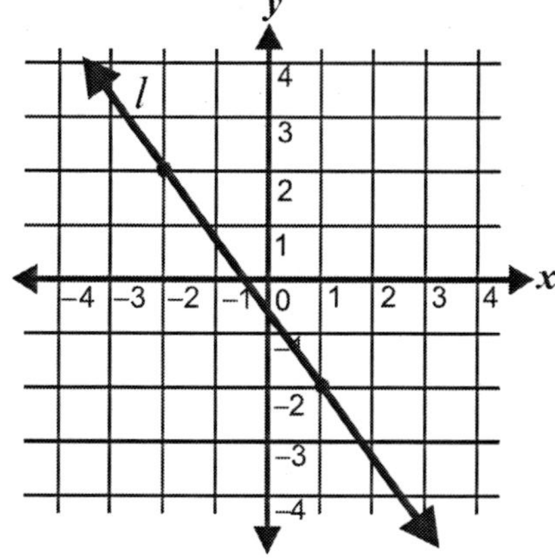

line l: ________________

line r: $y = \frac{1}{4}x - 2$

slopes: ________________

y-intercepts: ________________

15.

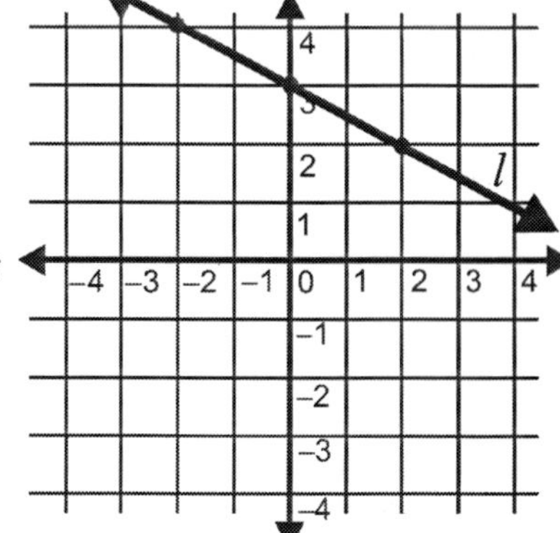

line l: ________________

line r: $y = -\frac{1}{2}x - 3$

slopes: ________________

y-intercepts: ________________

11.2 Equations of Perpendicular Lines

Now that we know how to calculate the slope of lines using two points, we are going to learn how to calculate the slope of a line perpendicular to a given line, then find the equation of that perpendicular line. To find the slope of a line perpendicular to any given line, take the slope of the first line, m:

1. multiply the slope by -1
2. invert (or flip over) the slope

You now have the slope of a perpendicular line. Writing the equation for a line perpendicular to another line involves three steps:

1. find the slope of the perpendicular line
2. choose one point on the first line
3. use the point-slope form to write the equation of the line perpendicular to the original line.

Example 3: The solid line on the graph below has a slope of $\frac{2}{3}$. Write the equation of a line perpendicular to the solid line.

Step 1: Find the slope of the perpendicular line. Multiply the slope by -1 and then flip it over (invert it).

$$\frac{2}{3} \times -1 = -\frac{2}{3} \curvearrowright -\frac{3}{2}$$

The slope of the perpendicular line, shown as a dotted line on the graph below, is $-\frac{3}{2}$.

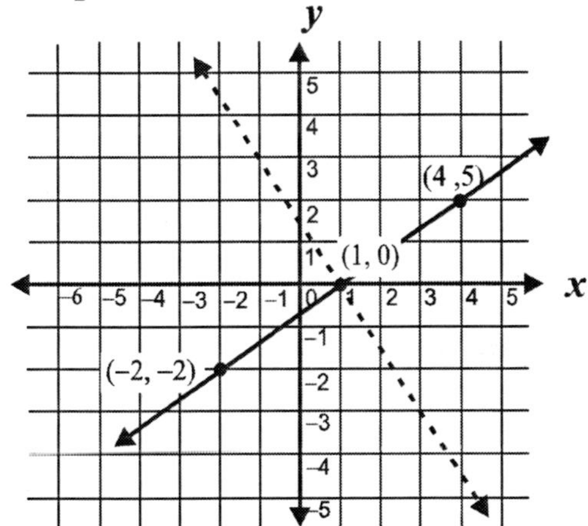

Step 2: Choose one point on the first line. We will use $(1, 0)$ in this example. We could also use the point $(-2, -2)$ or $(4, 5)$.

Step 3: Use the point-slope formula, $(y - y_1) = m(x - x_1)$, to write the equation of the perpendicular line. Remember, we chose $(1, 0)$ as our point. So, $(y - 0) = -\frac{3}{2}(x - 1)$. Simplified, $y = -\frac{3}{2}x + \frac{3}{2}$.

Solve the following problems involving perpendicular lines.

1. Find the slope of the line perpendicular to the solid line shown at right, and draw the perpendicular as a dotted line. Use one point on the solid line and the calculated slope to find the equation of the perpendicular line.

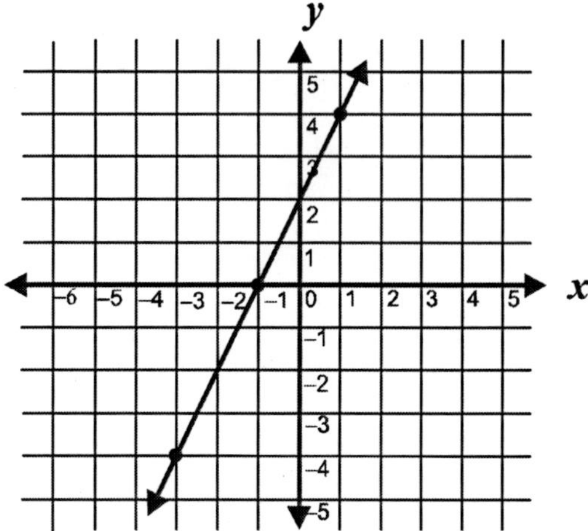

Find the equation of the perpendicular line using the point and slope given and the formula $(y - y_1) = m(x - x_1)$.

2. $(2, 1), 5$

3. $(3, 2), 2$

4. $(-2, 1), -3$

5. $(-4, 2), -\frac{1}{2}$

6. $(-1, 4), 1$

7. $(3, 3), \frac{2}{3}$

8. $(5, -1), -1$

9. $\left(\frac{1}{2}, \frac{3}{4}\right), 4$

10. $\left(\frac{2}{3}, \frac{3}{4}\right), -\frac{1}{6}$

11. $(7, -2), -\frac{1}{8}$

12. $(5, 0), \frac{4}{5}$

13. $(-3, -3), -\frac{7}{3}$

14. $\left(\frac{1}{4}, 4\right), \frac{1}{2}$

15. $(0, 6), -\frac{1}{9}$

11.3 Writing an Equation from Data

We often write data in a two-column format. If the increases or decreases in the ordered pairs are at a constant rate, then we can find a linear equation for the data.

Example 4: Write an equation for the following set of data.

Dan set his car on cruise control and noted the distance he went every 5 minutes.

Minutes in operation (x)	Odometer reading (y)
5	28,490 miles
10	28,494 miles

Step 1: Write two ordered pairs in the form (minutes, distance) for Dan's driving, $(5, 28490)$ and $(10, 28494)$, and find the slope.

$$m = \frac{28494 - 28490}{10 - 5} = \frac{4}{5}$$

Step 2: Use the ordered pairs to write the equation in the form $y = mx + b$. Place the slope, m, that you found and one of the pairs of points as x_1 and y_1 in the following formula, $y - y_1 = m(x - x_1)$.

$$y - 28490 = \tfrac{4}{5}(x - 5)$$
$$y - 28490 = \tfrac{4}{5}x - 4$$
$$y - 28490 + 28490 = \tfrac{4}{5}x - 4 + 28490$$
$$y + 0 = \tfrac{4}{5}x + 28486$$
$$y = \tfrac{4}{5}x + 28486$$

Write an equation for each of the following sets of data, assuming the relationship is linear.

1. **Doug's Doughnut Shop**

Year in Business	Total Sales
1	$55,000
4	$85,000

2. **Gwen's Green Beans**

Days Growing	Height in Inches
2	5
6	12

3. **Jim's Depreciation on His Jet Ski**

Years	Value
1	$4,500
6	$2,500

4. **Stepping on the Brakes**

Seconds	MPH
2	51
5	18

11.4 Graphing Linear Data

We relate many types of data by a constant ratio. As you learned on the previous page, this type of data is linear. The slope of the line described by linear data is the ratio between the data. Plotting linear data with a constant ratio can be helpful in finding additional values. Often linear data is not negative because many objects cannot be measured in negative values, and therefore, will not have any negative points. When there are no negative values for these type of problems, the actual graph will be a ray instead of a line.

Example 5: A department store prices socks per pair. Each pair of socks costs \$0.75. Plot pairs of socks versus price on a Cartesian plane.

Step 1: Since the price of the socks is constant, you know that one pair of socks costs \$0.75, 2 pairs of socks cost \$1.50, 3 pairs of socks cost \$2.25, and so on. Make a list of a few points.

Pair(s) x	Price y
1	0.75
2	1.50
3	2.25

Step 2: Plot these points on a Cartesian plane, and draw a straight line through the points.

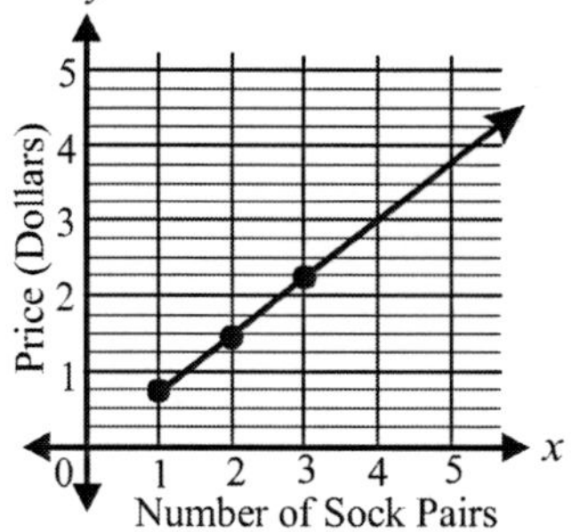

Example 6: What is the slope of the data in the example above? What does the slope describe?

Solution: You can determine the slope either by the graph or by the data points. For this data, the slope is 0.75. Remember, slope is rise/run. For every \$0.75 going up the y-axis, you go across one pair of socks on the x-axis. The slope describes the price per pair of socks.

Example 7: Use the graph created in the above example to answer the following questions. How much would 5 pairs of socks cost? How many pairs of socks could you purchase for \$3.00? Extending the line gives useful information about the price of additional pairs of socks.

Solution 1: The line that represents 5 pairs of socks intersects the data line at \$3.75 on the y-axis. Therefore, 5 pairs of socks would cost \$3.75.

Solution 2: The line representing the value of \$3.00 on the y-axis intersects the data line at 4 on the x-axis. Therefore, \$3.00 will buy exactly 4 pairs of socks.

Use the information given to make a line graph for each set of data, and answer the questions related to each graph.

1. The diameter of a circle versus the circumference of a circle is a constant ratio. Use the data given below to graph a line to fit the data. Extend the line, and use the graph to answer the next question.

Circle

Diameter	Circumference
4	12.56
5	15.70

2. Using the graph of the data in question 1, estimate the circumference of a circle that has a diameter of 3 inches.

3. If the circumference of a circle is 3 inches, about how long is the diameter?

4. What is the slope of the line you graphed in question 1?

5. What does the slope of the line in question 4 describe?

6. The length of a side on a square and the perimeter of a square are constant ratios to each other. Use the data below to graph this relationship.

Square

Length of side	Perimeter
2	8
3	12

7. Using the graph from question 6, what is the perimeter of a square with a side that measures 4 inches?

8. What is the slope of the line graphed in question 6?

9. Conversions are often constant ratios. For example, converting from pounds to ounces follows a constant ratio. Use the data below to graph a line that can be used to convert pounds to ounces.

Measurement Conversion

Pounds	Ounces
2	32
4	64

10. Use the graph from question 9 to convert 40 ounces to pounds.

11. What does the slope of the line graphed for question 9 represent?

12. Graph the data below, and create a line that shows the conversion from weeks to days.

Time

Weeks	Days
1	7
2	14

13. About how many days are in $2\frac{1}{2}$ weeks?

11.5 Identifying Graphs of Linear Equations

Match each equation below with the graph of the equation.

A. $x = -4$
B. $x = y$
C. $-\frac{1}{2}x = y$
D. $y = -4$
E. $4x + y = 4$
F. $y = x - 3$
G. $x - 2y = 6$
H. $2x + 3y = 6$
I. $y = 3x + 2$

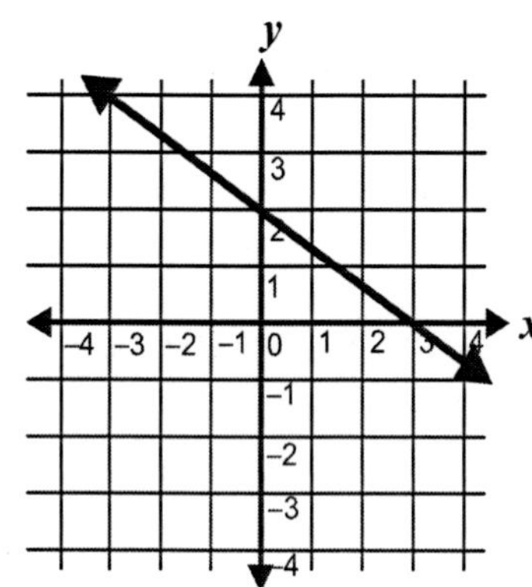

1. ________

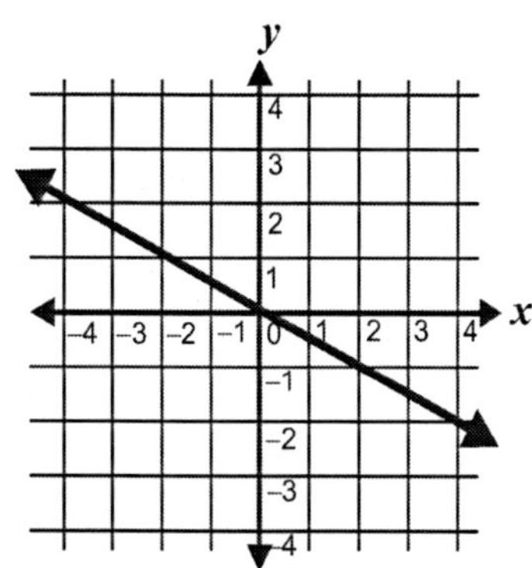

2. ________

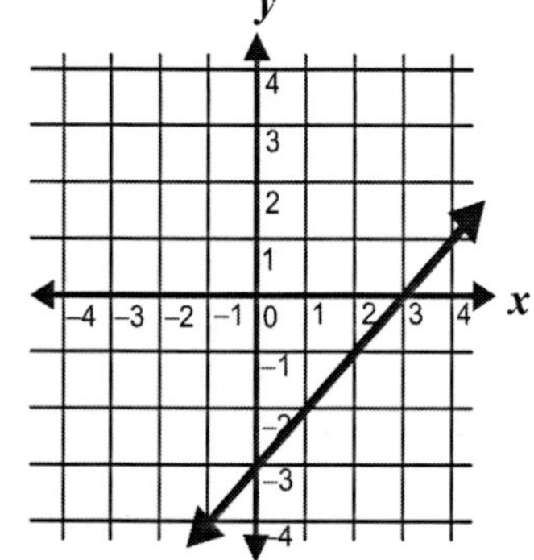

3. ________

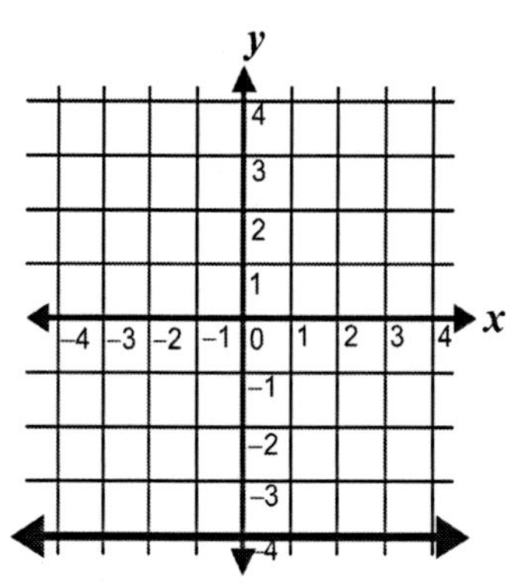

4. ________

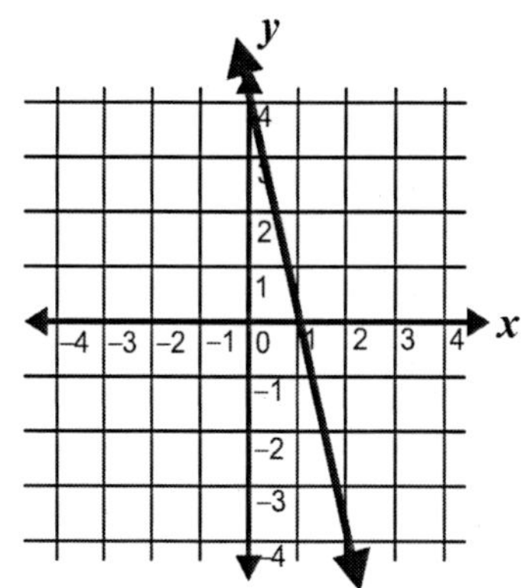

5. 

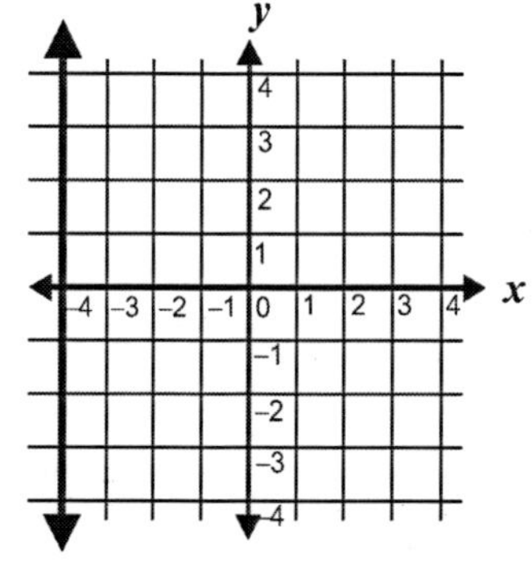

6. ________

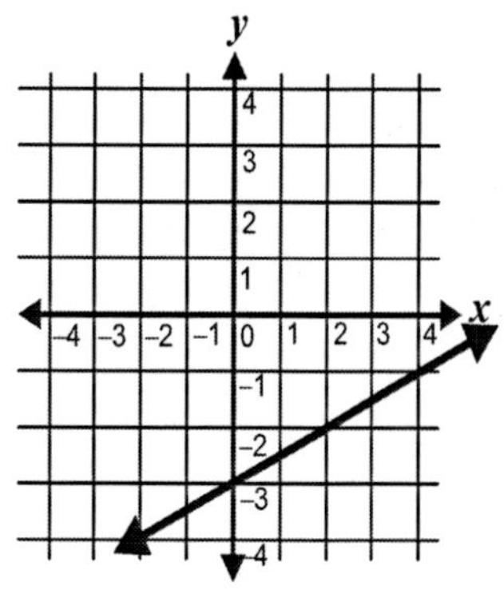

7. ________

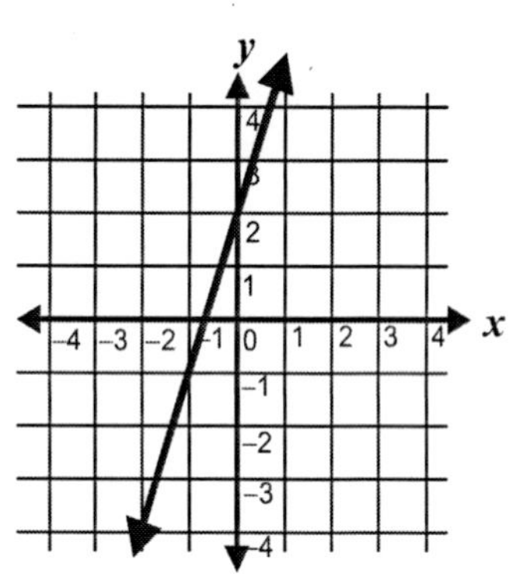

8. 

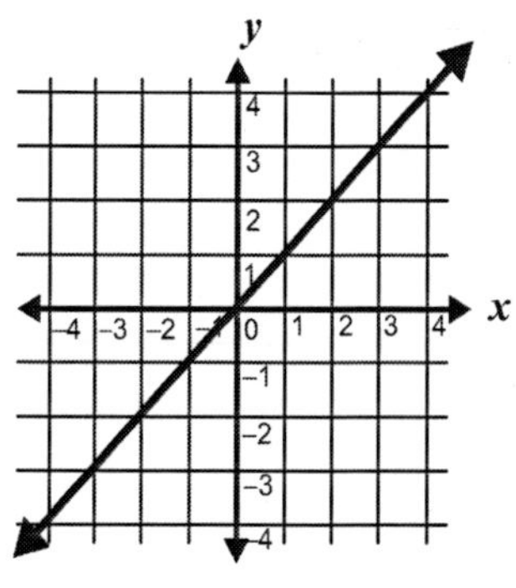

9. ________

11.6 Graphing Non-Linear Equations

Equations that you may encounter on the NC Algebra I End-of-Course exam may involve variables which are squared. The best way to find values for the x and y variables in an equation is to plug one number into x, and then find the corresponding value for y just as you did at the beginning of this chapter. Then, plot the points and draw a line through the points.

Example 8: Graph $y = x^2$.

Step 1: Make a table and find several values for x and y.

x	y
-2	4
-1	1
0	0
1	2
2	4

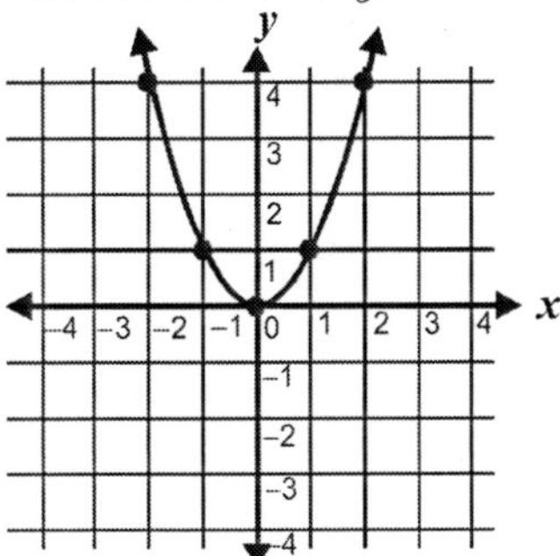

Step 2: Plot the points, and draw a curve through the points. Notice the shape of the curve. This type of curve is called a **parabola**.

Example 9: Graph the equation $y = -2x^2 + 4$.

Step 1: Make a table and find several values for x and y.

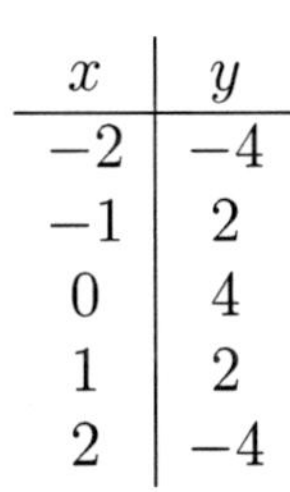

x	y
-2	-4
-1	2
0	4
1	2
2	-4

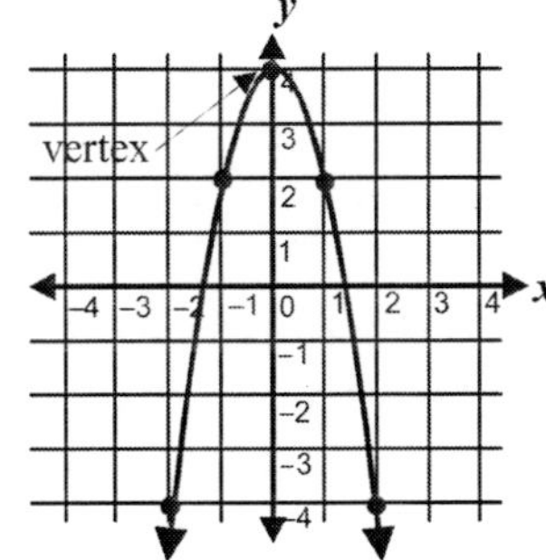

Step 2: Plot the points, and draw a curve through the points.

Note: In the equation $y = ax^2 + c$, changing the value of a will widen or narrow the parabola around the y-axis. If the value of a is a negative number, the parabola will be reflected across the x-axis (the vertex will be at the top of the parabola instead of at the bottom). If $a = 0$, the graph will be a straight line, not a parabola. Changing the value of c will move the vertex of the parabola from the origin to a different point on the y-axis.

Graph the equations below on a Cartesian plane.

1. $y = 2x^2$
2. $y = 3 - x^2$
3. $y = x^2 - 2$
4. $y = -2x^2$
5. $y = x^2 + 3$
6. $y = -3x^2 + 2$
7. $y = 3x^2 - 5$
8. $y = x^2 + 1$
9. $y = -x^2 - 6$
10. $y = -x^2$
11. $y = 2x^2 - 1$
12. $y = 2 - 2x^2$

11.7 Finding the Intercepts of a Quadratic Equation

A **quadratic equation** is an equation where either the y or x variable is squared. Finding the intercepts of a quadratic equation is similar to finding the intercepts of a line. In most cases, the variable x is squared, which means there will be two x-intercepts. The nonsquared variable will only have one intercept.

Example 10: Find the x-intercept(s) and the y-intercept(s) of the quadratic equation, $y = x^2 - 4$.

Step 1: First, find the y-intercept. Since the y-intercept is the point where the graph crosses the y-axis, the value for x at this point is zero. Because we know that $x = 0$, plug 0 in for x in the equation and solve for y.
$y = x^2 - 4$
$y = 0^2 - 4$
$y = 0 - 4$
$y = -4$
Therefore, the y-intercept is $(0, -4)$.

Step 2: Next, find the x-intercept(s). The x-intercept is the point, or points in this case, where the graph crosses the x-axis. In this case, we plug 0 in for y because y is always zero along the x-axis. Solve for x.
$y = x^2 - 4$
$0 = x^2 - 4$
$0 + 4 = x^2 - 4 + 4$
$4 = x^2$
$\sqrt{4} = \sqrt{x^2}$
$\sqrt{4} = x$
Thus, $x = \sqrt{4}$ and $\sqrt{4} = -2$ or 2, so $x = -2$ or $x = 2$.
The x-intercepts are $(-2, 0)$ and $(2, 0)$.

Step 3: To verify that the intercepts are correct, graph the equation on the coordinate plane.

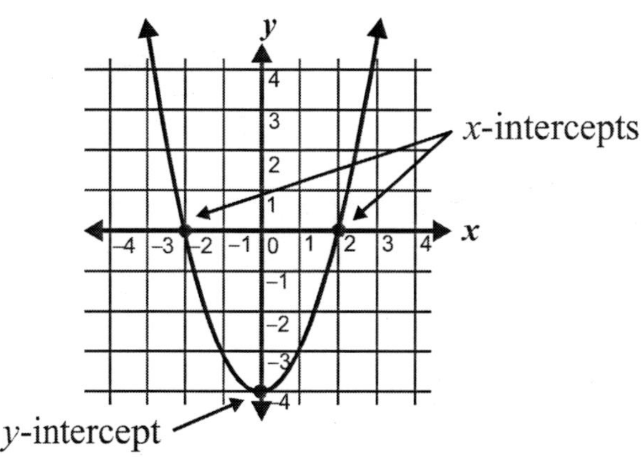

Example 11: Find the x-intercept and the y-intercept of the quadratic equation, $y = x^2 + 2$.

Step 1: First, find the y-intercept. Plug 0 in for x in the equation and solve for y.
$y = x^2 + 2 = 0^2 + 2 = 0 + 2 = 2$
Therefore, the y-intercept is $(0, 2)$.

Step 2: Next, find the x-intercept. Plug 0 in for y and solve for x.
$y = x^2 + 2$
$0 = x^2 + 2$
$0 - 2 = x^2 + 2 - 2$
$-2 = x^2$
$\sqrt{-2} = \sqrt{x^2}$
$\sqrt{-2} = x$
You cannot take the square root of a negative number and get a real number as the answer, so there is no x-intercept for this quadratic equation.

Step 3: To verify that the intercepts are correct, graph the equation on the coordinate plane. As you can see from the graph, the tails of the parabola are increasing in the positive y direction, so they will never come back down, which means they will never cross the x-axis.

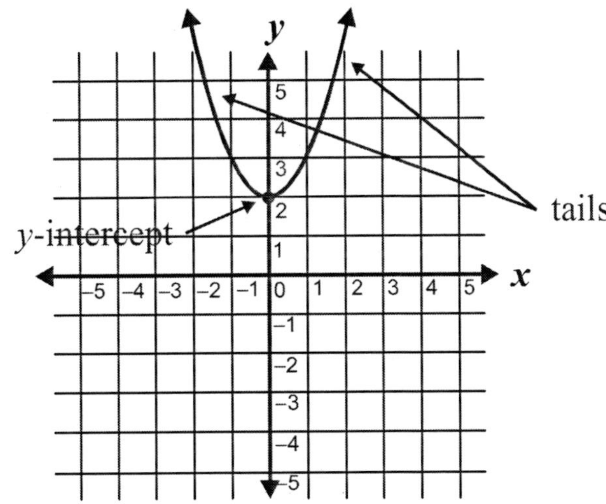

Find the x-intercept and y-intercept of the following quadratic equations.

1. $y = x^2$
2. $y = 2x^2 - 4$
3. $y = -x^2 + 2$
4. $y = x^2 + 6$
5. $y = -x^2 - 1$
6. $y = x^2 - 5$
7. $y = 4x^2 + 8$
8. $y = x^2 + 2x + 1$
9. $y = x^2 - 7x + 12$

11.8 Finding the Vertex of a Quadratic Equation

The vertex of a quadratic equation is the point on the parabola where the graph changes directions from increasing to decreasing or decreasing to increasing.

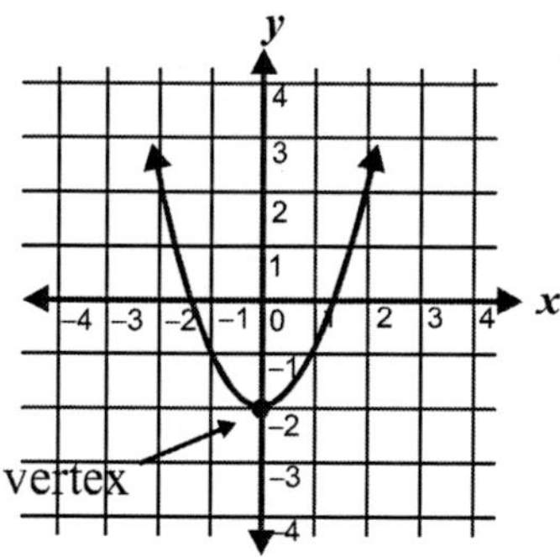

As a reminder, the quadratic equation is defined as $y = ax^2 + bx + c$, in which the coefficient a can never equal zero. If $a = 0$, then the equation will not have a x^2 term, which is what makes it a quadratic equation. The quadratic equation can also be written as a function of x by substituting $f(x)$ for y, such as $f(x) = ax^2 + bx + c$. To find the point of the vertex of the graph, you must use the formula below.

$$\text{vertex} = \left(-\frac{b}{2a}, f\left(-\frac{b}{2a}\right)\right)$$

where $f\left(-\frac{b}{2a}\right)$ is the quadratic equation evaluated at the value $-\frac{b}{2a}$. To do this, plug $-\frac{b}{2a}$ in for x.

Example 12: Find the point of the vertex of $f(x) = 2x^2 - 12x + 10$.

Step 1: First, find out what a and b equal. Since a is the coefficient of x^2, $a = 2$. b is the coefficient of x, so $b = -12$.

Step 2: Find the solution to $-\frac{b}{2a}$ by substituting the values of a and b from the equation into the expression.

$$-\frac{b}{2a} = -\left(\frac{-12}{2 \times 2}\right) = -\left(\frac{-12}{4}\right) = -(-3) = 3$$

Step 3: Find the solution to $f\left(-\frac{b}{2a}\right)$. We know that $-\frac{b}{2a} = 3$, so we need to find $f(3)$. To do this, we must substitute 3 into the quadratic equation for x.
$f(x) = 2x^2 - 12x + 10$
$f(3) = 2(3)^2 - 12(3) + 10 = 2(9) - 36 + 10 = 18 - 36 + 10 = -8$
The vertex equals $(3, -8)$.

Find the point of the vertex for the following equations.

1. $f(x) = x^2 + 6$
2. $f(x) = 2x^2 - 8$
3. $f(x) = x^2 + 10x - 4$
4. $f(x) = 2x^2 - 16x - 8$
5. $f(x) = x^2 + 4x - 5$
6. $f(x) = 3x^2 - 12x + 16$
7. $f(x) = x^2 - 25$
8. $f(x) = -x^2 + 6x - 12$
9. $f(x) = 4x^2 - 64x + 200$

11.9 Identifying Graphs of Real-World Situations

Real-world situations are sometimes modeled by graphs. Although an equation cannot be written for most of these graphs, interpreting these graphs provides valuable information. Situations may be represented on a graph as a function of time, length, temperature, etc.

The graph below depicts the temperature of a pond at different times of the day. Refer to the graph as you read through examples 13 and 14.

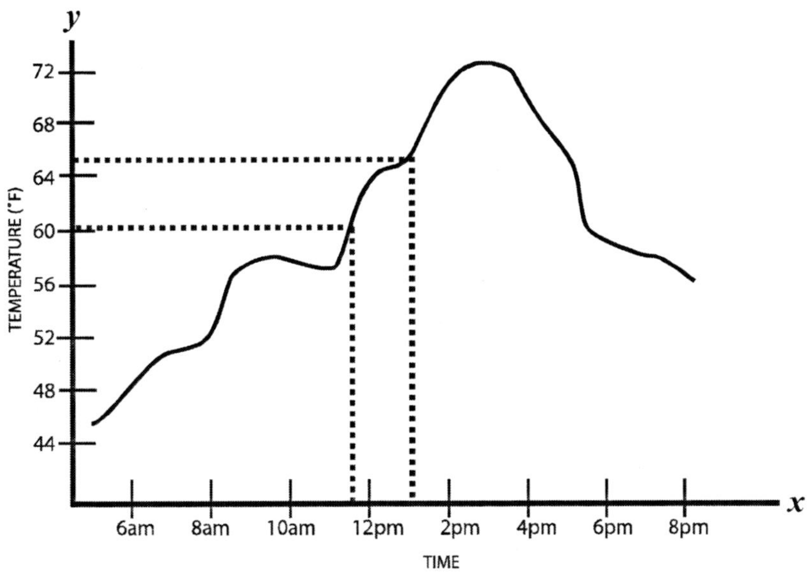

Example 13: If it is known that a specific breed of fish is most active in waters between 60°F and 65°F, what time of the day would this fish be the most active in this particular pond?

To find the answer, draw lines from the 60°F and 65°F points on the y-axis to the graph. Then, draw vertical lines from the graph to the x-axis. The time range between the two vertical lines on the x-axis indicates the time that the fish are most active. It can be determined from the graph that the fish are most active between 11:30 AM and 1:00 PM.

Example 14: Describe the way the temperature of the pond acts as a function of time.

At 6 AM, the temperature of the pond is about 47°F. The temperature increases relatively steadily throughout the morning and early afternoon. The temperature peaks at 72°F, which is around 2:30 PM during the day. Afterwards, the temperature of the pond starts to decrease. The later it gets in the evening, the more the temperature of the water decreases. The graph shows that at 8 PM the temperature of the pond is about 57°F.

Use the graphs to answer the questions. Circle your answers.

The following graph depicts the number of articles of clothing as a function of time throughout the year. Use this graph for questions 1 and 2.

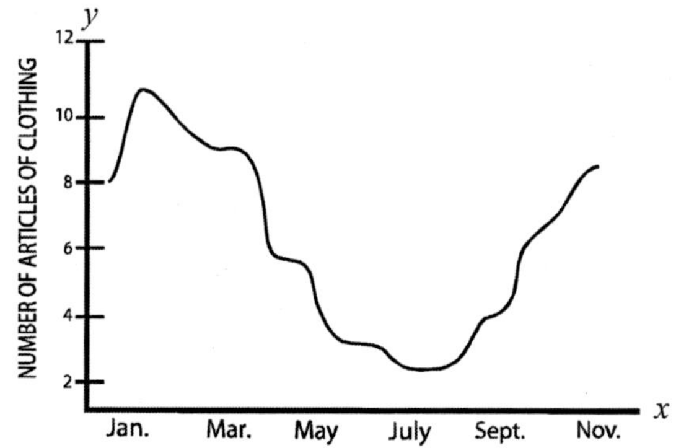

1. According to the graph, in what month are the most articles of clothing worn?
   (A) January
   (B) March
   (C) May
   (D) November

2. What is the average number of clothes a person wears in June?
   (A) 6
   (B) 3
   (C) 2
   (D) 5

The graph below depicts the efficiency of energy transfer as a function of distance in a certain element. Use the graph to answer questions 3 and 4.

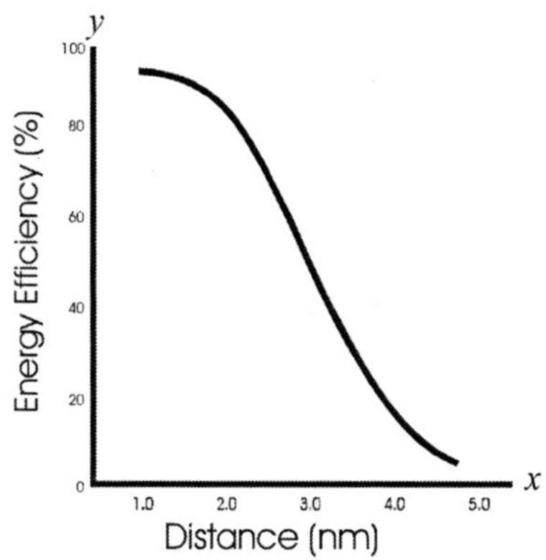

3. At what distance is the energy efficiency at 50%?
   (A) 1.0 nm
   (B) 2.0 nm
   (C) 3.5 nm
   (D) 3.0 nm

4. What is the energy efficiency at distance 2.5 nm?
   (A) 100%
   (B) 90%
   (C) 80%
   (D) 70%

Find the best non-linear graph to match each scenario.

1. Cathy begins the two-hour drive to her mother's house in her new sedan. She drives slowly through her city for thirty minutes to reach Interstate 95. After she enters the highway, she travels a constant 60–70 miles per hour for the next hour until she reaches her mother's exit. She then drives slowly down back roads to arrive at her mother's house.

(A)

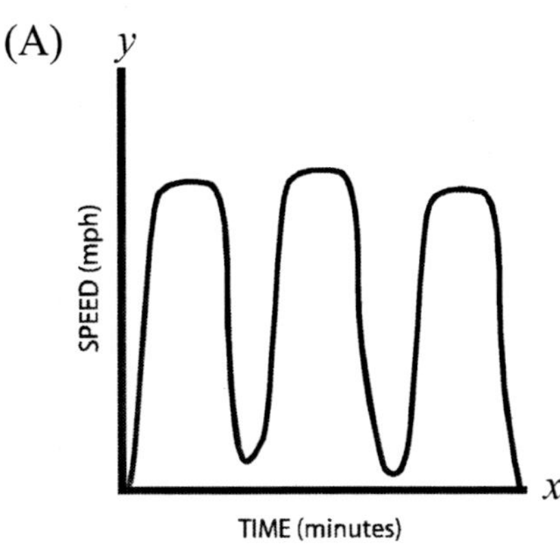

2. Phillip is flying to Texas for a business meeting. When his flight leaves, the airplane increases its speed a great deal until it reaches about 550 miles per hour. After 20 minutes, the plane levels off for the last 45 minutes at 500 miles per hour. As the airplane nears the airport in Fort Worth, TX, it decreases its speed until it lands and reaches zero miles per hour.

(B)

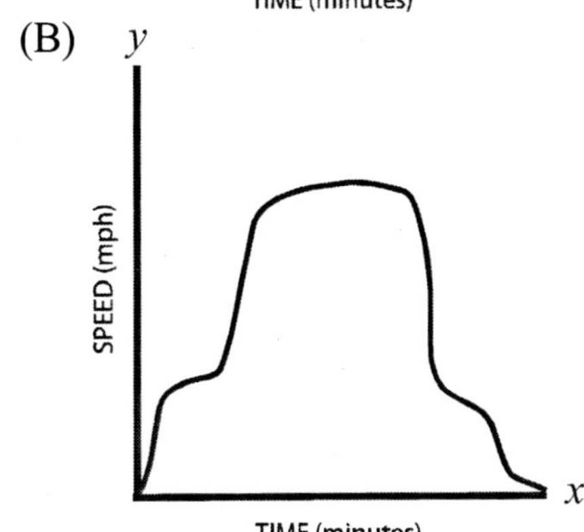

3. Erica and her father like to build rockets for fun, and every Saturday they go to the park by their house to launch the rockets. Almost immediately after takeoff, the rocket reaches its greatest speed. Affected by gravity, it slows down until it reaches its peak height. It again speeds up as it descends to the ground.

(C)

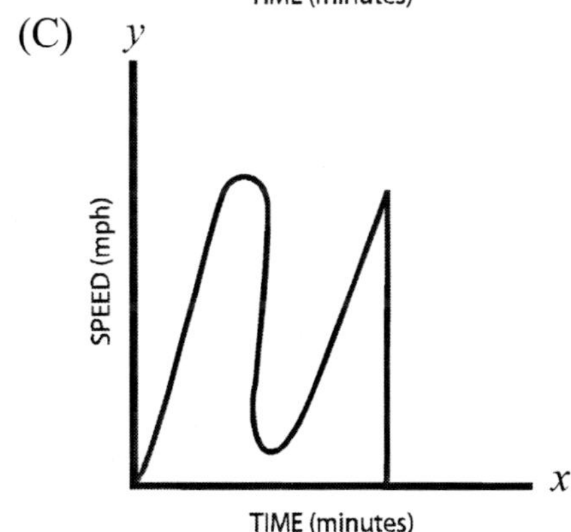

4. Molly and her mother ride the train each time they go to the zoo. Molly knows that the train slows down twice so that the passengers can view the animals. Her favorite part of the ride, though, is when the train moves very quickly before it slows down to approach the station and come to a stop.

(D)

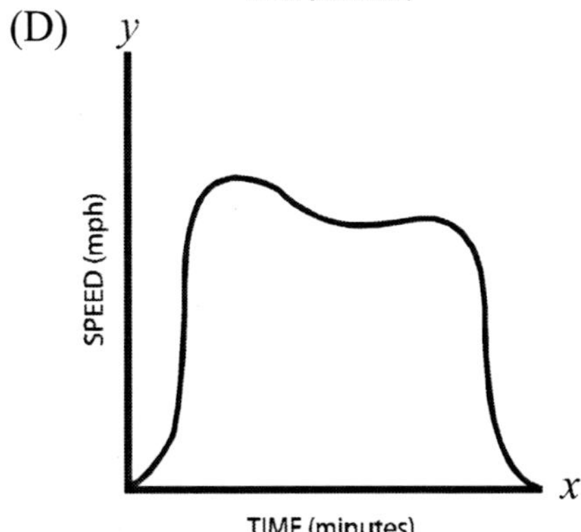

11.10 Scatter Plots

A **scatter plot** is a graph of ordered pairs involving two sets of data. These plots are used to detect whether two sets of data, or variables, are truly related.

In the example to the right, two variables, income and education, are being compared to see if they are related or not. Twenty people were interviewed, ages 25 and older, and the results were recorded on the chart.

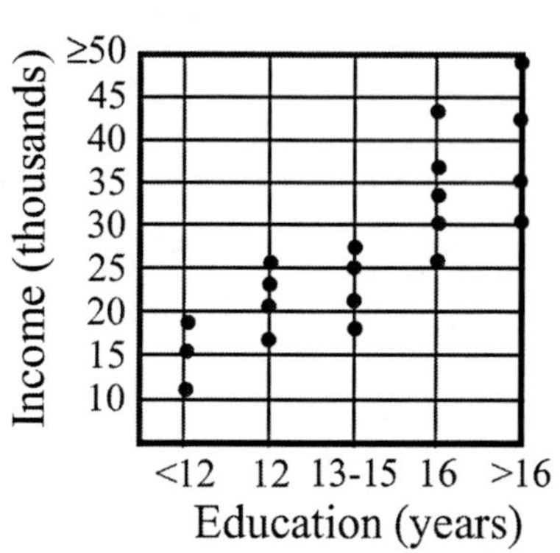

Imagine drawing a line on the scatter plot where half of the points are above the line and half the points are below it. In the plot on the right, you will notice that this line slants upward and to the right. This line direction means there is a **positive** relationship between education and income. In general, for every increase in education, there is a corresponding increase in income.

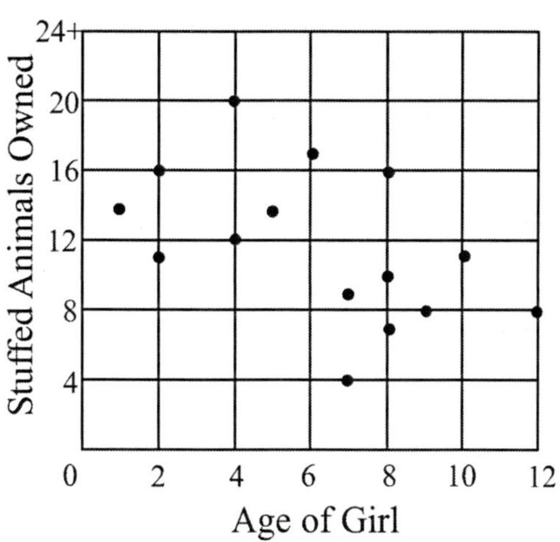

Now, examine the scatter plot on the left. In this case, 15 girls ages 2–12 were interviewed and asked, "How many stuffed animals do you currently have?" If you draw an imaginary line through the middle points, you will notice that the line slants downward and to the right. This plot demonstrates a **negative** relationship between the ages of girls and their stuffed animal ownership. In general, as the girls' ages increase, the number of stuffed animals owned decreases.

Finally, look at the scatter plot shown on the right. In this plot, Rita wanted to see the relationship between the temperature in the classroom and the grades she received on tests she took at that temperature. As you look to your right, you will notice that the points are distributed all over the graph. Because this plot is not in a pattern, there is no way to draw a line through the middle of the points. This type of point pattern indicates there is **no relationship** between Rita's grades on tests and the classroom temperature.

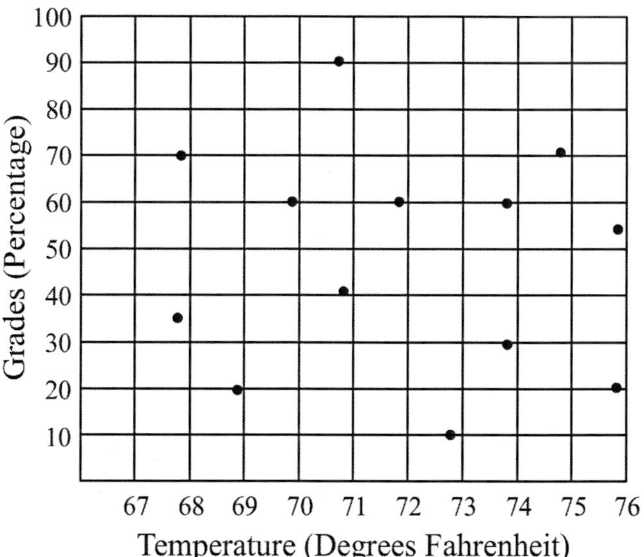

Examine each of the scatter plots below. Write whether the relationship shown between the two variables is "positive," "negative," or "no relationship."

1.

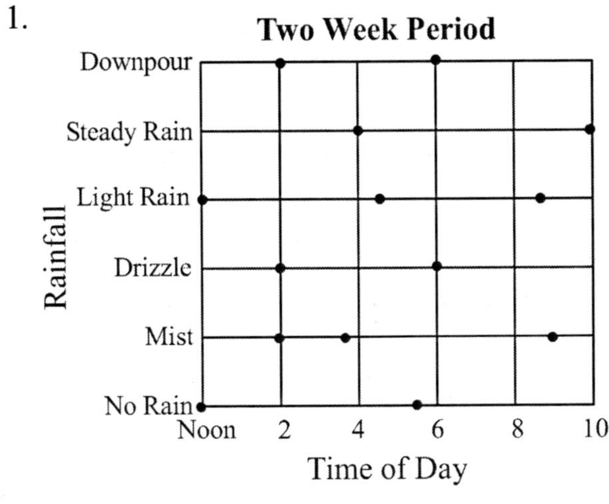

4.

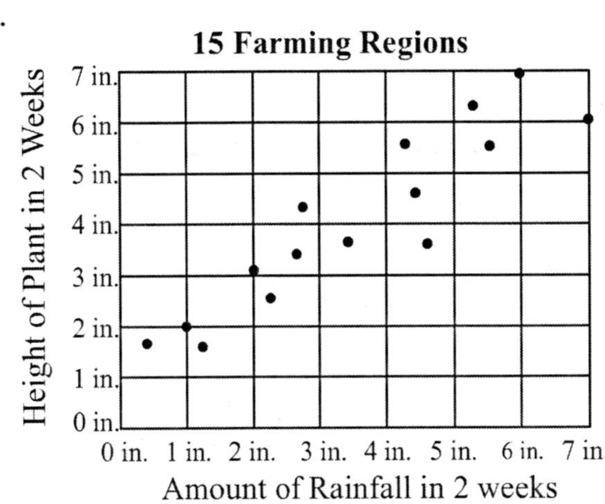

2.

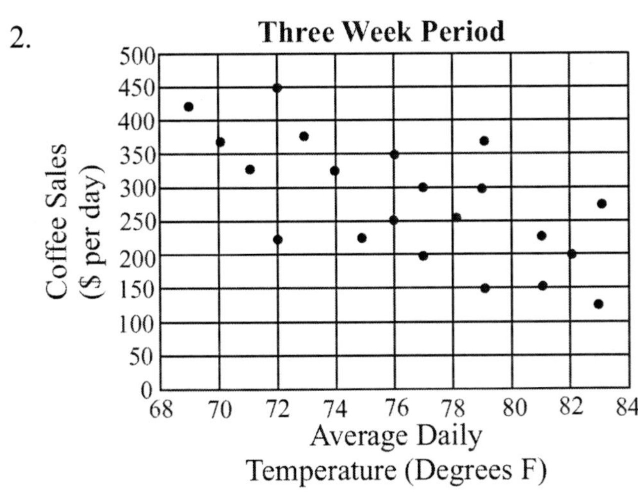

5.

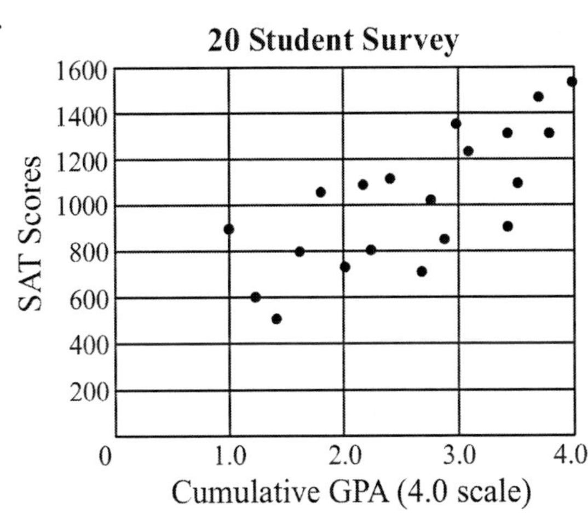

3.

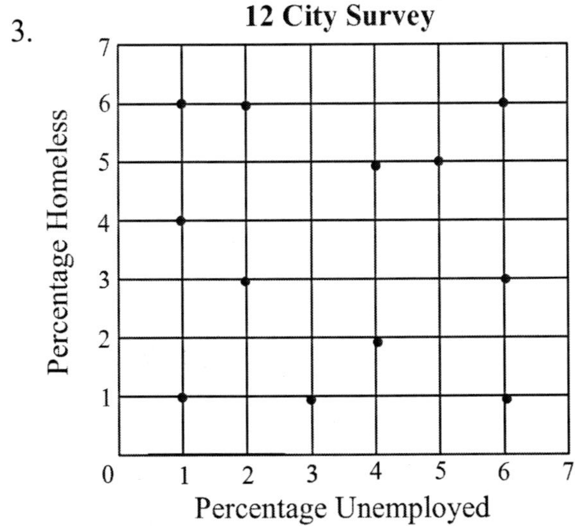

6.

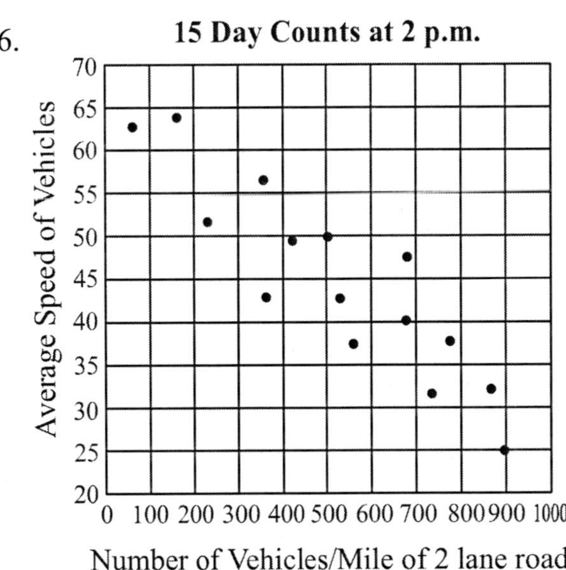

11.11 The Line of Best Fit

At this point, you now understand how to plot points on a Cartesian plane. You also understand how to find the data trend on a Cartesian plane. These skills are necessary to accomplish the next task, determining the line of best fit.

In order to find the line of best fit, you must first draw a scatter plot of all data points. Once this is accomplished, draw an oval around all of the points plotted. Draw a line through the points in such a way that the line separates half the points from one another. You may now use this line to answer questions.

Example 15: The following data set contains the heights of children between 5 and 13 years old. Make a scatter plot and draw the line of best fit to represent the trend. Using the graph, determine the height for a 14-year old child.

Age 5: 4'6", 4'4", 4'5"	Age 8: 4'8", 4'6", 4'7"	Age 11: 5'0", 4'10"
Age 6: 4'7", 4'5", 4'6"	Age 9: 4'9", 4'7", 4'10"	Age 12: 5'1", 4'11", 5'0", 5'3"
Age 7: 4'9", 4'7", 4'6", 4'8"	Age 10: 4'9", 4'8", 4'10"	Age 13: 5'3", 5'2", 5'0", 5'1"

In this example, the data points lay in a positive sloping direction. To determine the line of best fit, all data points were circled, then a line of best fit was drawn. Half of the points lay below, half above the line of best fit drawn bisecting the narrow length of the oval.

To find the height of a 14-year old, simply continue the line of best fit forward. In this case, the height is 62 inches.

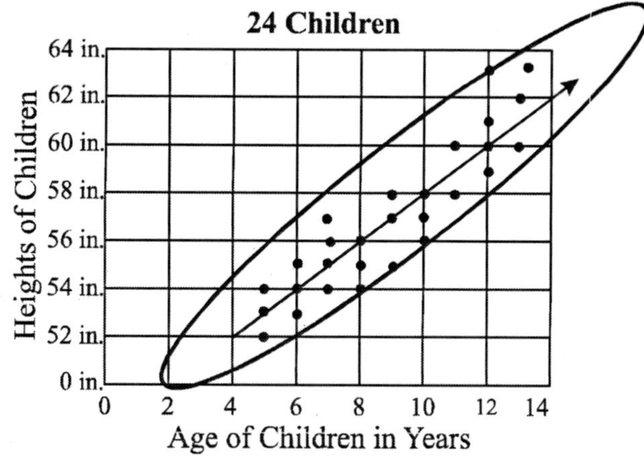

Plot the data sets below, then draw the line of best fit. Next, use the line to estimate the value of the next measurement.

1. Selected values of the Sleekster Brand Light Compact Vehicles: New Vehicle: $13, 000.
   1 year old: $12, 000, $11, 000, $12, 500   3 year old: $8, 500, $8, 000, $9, 000
   2 year old: $9, 000, $10, 500, $9, 500   4 year old: $7, 500, $6, 500, $6, 000
   5 year old: ?

2. The relationship between string length and kite height for the following kites:
   (L = 500 ft, H = 400 ft) (L = 250 ft, H = 150 ft) (L = 100 ft, H = 75 ft) (L = 500 ft, H = 350 ft) (L = 250 ft, H = 200 ft) (L = 100 ft, H = 50 ft) (L = 600 ft, H = ?)

3. Relationship between Household Incomes(HI) and Household Property Values (HPV):
   (HI = $30,000, HPV = $100,000) (HI = $45,000, HPV = $120,000) (HI = $60,000, HPV = $135,000) (HI = $50,000, HPV = 115,000) (HI = $35,000, HPV = 105,000) (HI = 65,000, HPV = 155,000) (HI = $90,000, HPV = ?)

Chapter 11 Review

1. Paulo turns on the oven to preheat it. After one minute, the oven temperature is 200°. After 2 minutes, the oven temperature is 325°.

Oven Temperature

Minutes	Temperature
1	200°
2	325°

Assuming the oven temperature rises at a constant rate, write an equation that fits the data.

2. Write an equation that fits the data given below. Assume the data is linear.

Plumber Charges per Hour

Hour	Charge
1	\$170
2	\$220

3. What is the name of the curve described by the equation $y = 2x^2 - 1$?

4. Graph the equation $y = -\frac{1}{2}x^2 + 1$ on your own graph paper.

5. What happens to a graph if the slope changes from 2 to -2?

(A) The graph will move down 4 spaces.
(B) The graph will slant downward towards the left instead of the right.
(C) The graph will flatten out to be more vertical.
(D) The graph will slant downward towards the right instead of the left.

6. What happens to a graph if the y-intercept changes from 4 to -2?

(A) The graph will move down 2 spaces.
(B) The graph will slant towards the left instead of the right.
(C) The graph will move down 6 spaces.
(D) The graph will move up 6 spaces.

7. The graph of the line $y = 3x - 1$ is shown below. On the same graph, draw the line $y = -\frac{1}{3}x - 1$.

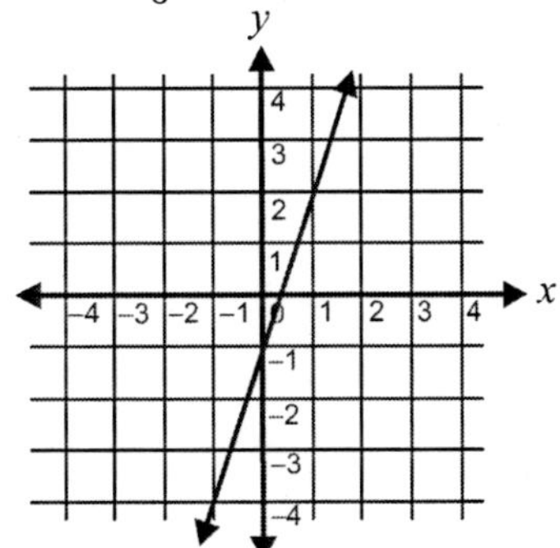

8. Which of the following statements is an accurate comparison of the lines $y = 3x - 1$ and $y = -\frac{1}{3}x - 1$?

(A) Only their y-intercepts are different.
(B) Only their slopes are different.
(C) Both their y-intercepts and their slopes are different.
(D) There is no difference between these two lines.

9. The data given below show conversions between miles per hour and kilometers per hour. Based on this data, graph a conversion line on the Cartesian plane below.

Speed

MPH	KPH
5	8
10	16

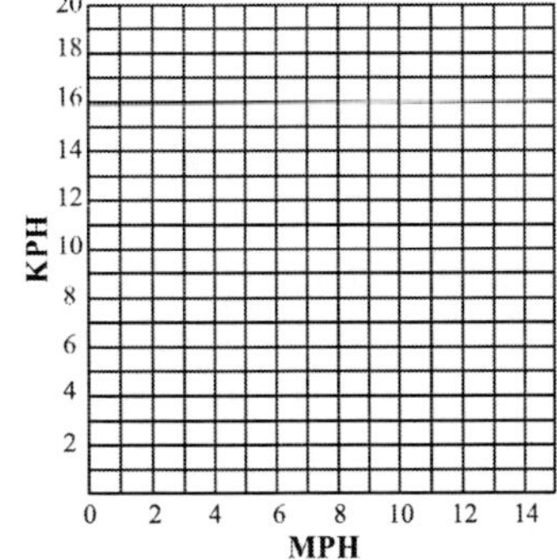

10. What would be the approximate conversion of 9 mph to kph?

11. What would be the approximate conversion of 13 kph to mph?

12. A bicyclist travels 12 mph downhill. Approximately how many kph is the bicyclist traveling?

13. Use the data given below to graph the interest rate versus the interest rate on \$80.00 in one year.

\$80.00 Principal

Interest Rate	Interest-1 year
5%	\$4.00
10%	\$8.00

14. About how much interest would accrue in one year at an 8% interest rate?

15. What is the slope of the line describing interest versus interest rate?

16. What information does the slope give in problem 15?

17. Draw the graph of the following situation on the Cartesian plane provided. A girl rode her bicycle up a hill, then coasted down the other side of the hill on her bike. At the bottom she stopped.

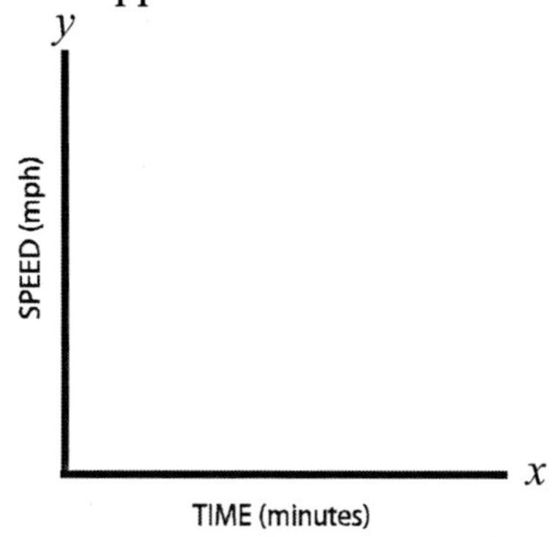

18. Find the x-intercept(s) and y-intercept of $y = -3x^2$.

19. Find the x-intercept(s) and y-intercept of $y = -4x^2 + 16$.

20. Find the point of the vertex of $f(x) = 2x^2 - 8x + 15$

For questions 21 and 22, write whether the relationship shown between the two variables is "positive," "negative," or "no relationship."

21.

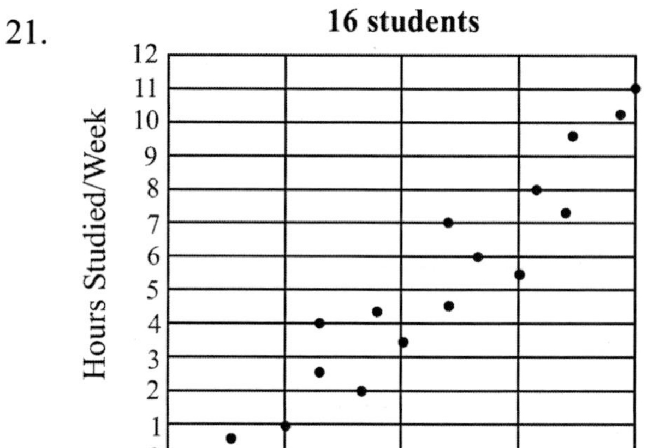

22.

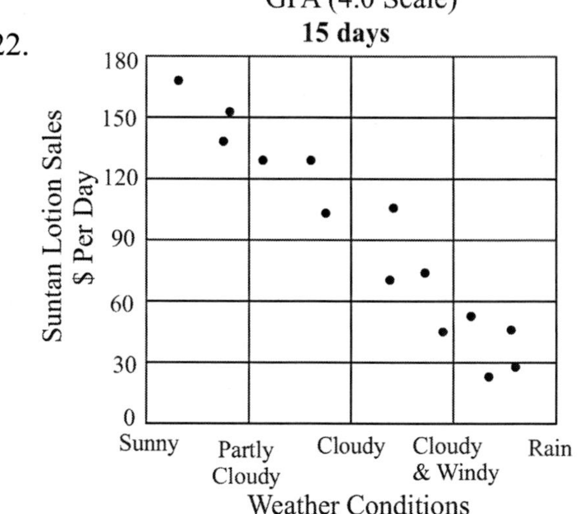

Chapter 12
Systems of Equations and Systems of Inequalities

This chapter covers the following North Carolina mathematics standards for Algebra I:

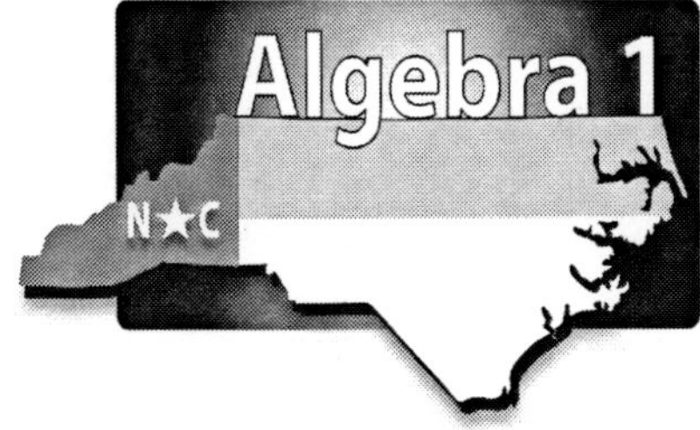

Competency Goal	Objectives
Number and Operations	1.01 1.02
Geometry and Measurement	2.02
Algebra	4.01a 4.03

We call two linear equations considered at the same time a **system** of linear equations. The graph of a linear equation is a straight line. The graphs of two linear equations can show that the lines are **parallel**, **intersecting**, or **collinear**. Two lines that are **parallel** will never intersect and have no ordered pairs in common. If two lines are **intersecting**, they have one point in common, and in this chapter, you will learn to find the ordered pair for that point. If the graph of two linear equations is the same line, we say the lines are **collinear**.

If you are given a system of two linear equations, and you put both equations in slope-intercept form, you can immediately tell if the graph of the lines will be **parallel**, **intersecting**, or **collinear**.

If two linear equations have the same slope and the same y-intercept, then they are both equations for the same line. They are called **collinear** or **coinciding** lines. A line is made up of an infinite number of points extending infinitely far in two directions. Therefore, collinear lines have an infinite number of points in common.

Example 1: $2x + 3y = -3$ **In slope intercept form:** $y = -\frac{2}{3}x - 1$

$4x + 6y = -6$ **In slope intercept form:** $y = -\frac{2}{3}x - 1$

The slope and y-intercept of both lines are the same.

If two linear equations have the same slope but different y-intercepts, they are **parallel** lines. Parallel lines never touch each other, so they have no points in common.

If two linear equations have different slopes, then they are intersecting lines and share exactly one point in common.

The chart below summarizes what we know about the graphs of two equations in slope-intercept form.

y-Intercepts	Slopes	Graphs	Number of Solutions
same	same	collinear	infinite
different	same	distinct parallel lines	none (they never touch)
same or different	different	intersecting lines	exactly one

For the pairs of equations below, put each equation in slope-intercept form, and tell whether the graphs of the lines will be collinear, parallel, or intersecting.

1. $x - y = -1$
   $-x + y = -1$

2. $x - 2y = 4$
   $-x + 2y = 6$

3. $y - 2 = x$
   $x + 2 = y$

4. $x = y - 1$
   $-x = y - 1$

5. $2x + 5y = 10$
   $4x + 10y = 20$

6. $x + y = 3$
   $x - y = 1$

7. $2y = 4y - 6$
   $-6x + y = 3$

8. $x + y = 5$
   $2x + 2y = 10$

9. $2x = 3y - 6$
   $4x = 6y - 6$

10. $2x - 2 = 2$
    $3y = -x + 5$

11. $x = -y$
    $x = 4 - y$

12. $2x = y$
    $x + y = 3$

13. $x = y + 1$
    $y = x + 1$

14. $x - 2y = 4$
    $-2x + 4y = -8$

15. $2x + 3y = 4$
    $-2x + 3y = -8$

16. $2x - 4y = 1$
    $-6x + 12y = 3$

17. $-3x + 4y = 1$
    $6x + 8y = 2$

18. $x + y = 2$
    $5x + 5y = 10$

19. $x + y = 4$
    $x - y = 4$

20. $y = -x + 3$
    $x - y = 1$

12.1 Finding Common Solutions for Intersecting Lines

When two lines intersect, they share exactly one point in common.

Example 2: $3x + 4y = 20$ and $2y - 4x = 12$

Put each equation in slope-intercept form.

$3x + 4y = 20$	$2y - 4x = 12$
$4y = -3x + 20$	$2y = 4x + 12$
$y = -\frac{3}{4}x + 5$	$y = 2x + 6$

slope-intercept form

Straight lines with different slopes are **intersecting lines**. Look at the graphs of the lines on the same Cartesian plane.

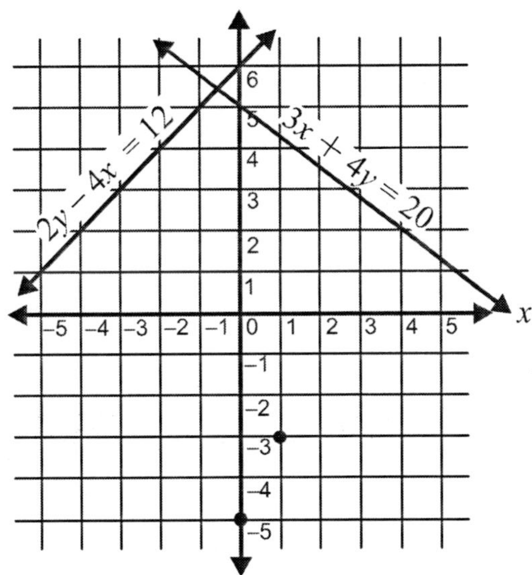

You can see from looking at the graph that the intersecting lines share one point in common. However, it is hard to tell from looking at the graph what the coordinates are for the point of intersection. To find the exact point of intersection, you can use the **substitution method** to solve the system of equations algebraically.

12.2 Solving Systems of Equations by Substitution

You can solve systems of equations by using the substitution method.

Example 3: Find the point of intersection of the following two equations:

Equation 1: $x - y = 3$

Equation 2: $2x + y = 9$

Step 1: Solve one of the equations for x or y. Let's choose to solve equation 1 for x.

Equation 1: $x - y = 3$

$x = y + 3$

Step 2: Substitute the value of x from equation 1 in place of x in equation 2.

Equation 2:
$$\begin{aligned} 2x + y &= 9 \\ 2(y+3) + y &= 9 \\ 2y + 6 + y &= 9 \\ 3y + 6 &= 9 \\ 3y &= 3 \\ y &= 1 \end{aligned}$$

Step 3: Substitute the solution for y back in equation 1 and solve for x.

Equation 1:
$$\begin{aligned} x - y &= 3 \\ x - 1 &= 3 \\ x &= 4 \end{aligned}$$

Step 4: The solution set is $(4, 1)$. Substitute $(4, 1)$ in both of the equations to check.

Equation 1:
$$\begin{aligned} x - y &= 3 \\ 4 - 1 &= 3 \\ 3 &= 3 \end{aligned}$$

Equation 2:
$$\begin{aligned} 2x + 9 &= 9 \\ 2(4) + 1 &= 9 \\ 8 + 1 &= 9 \\ 9 &= 9 \end{aligned}$$

The point $(4, 1)$ is common for both equations. This is the **point of intersection**.

For each of the following pairs of equations, find the point of intersection, the common solution, using the substitution method.

1. $x+2y=8$
   $2x-3y=2$
2. $x-y=-5$
   $x+y=1$
3. $x-y=4$
   $x+y=2$
4. $x-y=-1$
   $x+y=9$
5. $-x+y=2$
   $x+y=8$
6. $x+4y=10$
   $x+5y=10$
7. $2x+3y=2$
   $4x-9y=-1$
8. $x+3y=5$
   $x-y=1$
9. $-x=y-1$
   $x=y-1$
10. $x-2y=2$
    $2y+x=-2$
11. $5x+2y=1$
    $2x+4y=10$
12. $3x-y=2$
    $5x+y=6$
13. $2x+3y=3$
    $4x+5y=5$
14. $x-y=1$
    $-x-y=1$
15. $x=y+3$
    $y=3-x$

12.3 Solving Systems of Equations by Adding or Subtracting

You can solve systems of equations algebraically by adding or subtracting an equation from another equation or system of equations.

Example 4: Find the point of intersection of the following two equations:
Equation 1: $x+y=10$
Equation 2: $-x+4y=5$

Step 1: Eliminate one of the variables by adding the two equations together. Since the x has the same coefficient in each equation, but opposite signs, it will cancel nicely by adding.

$$\begin{array}{r} x+y=10 \\ +(-x+4y=5) \\ \hline 0+5y=15 \\ 5y=15 \\ y=3 \end{array}$$

Add each like term together.
Simplify.
Divide both sides by 5.

Step 2: Substitute the solution for y back into an equation, and solve for x.
Equation 1: $x+y=10$ Substitute 3 for y.
$x+3=10$ Subtract 3 from both sides.
$x=7$

Step 3: The solution set is $(7,3)$. To check, substitute the solution into both of the original equations.

Equation 1:
$$\begin{aligned} x+y&=10 \\ 7+3&=10 \\ 10&=10 \end{aligned}$$

Equation 2:
$$\begin{aligned} -x+4y&=5 \\ -(7)+4(3)&=5 \\ -7+12&=5 \\ 5&=5 \end{aligned}$$

The point $(7,3)$ is the point of intersection.

Example 5: Find the point of intersection of the following two equations:
Equation 1: $3x - 2y = -1$
Equation 2: $-4y = -x - 7$

Step 1: Put the variables in equation 2 on the same side.

$-4y = -x - 7$ Add x to both sides.
$x - 4y = -x + x - 7$ Simplify.
$x - 4y = -7$

Step 2: Add the two equations together to cancel one variable. Since each variable has the same sign and different coefficients, we have to multiply one equation by a negative number so one of the variables will cancel. Equation 1's y variable has a coefficient of 2, and if multiplied by -2, the y will have the same variable as the y in equation 2, but a different sign. This will cancel nicely when added.

$-2\,(3x - 2y = -1)$ Multiply by -2.
$-6x + 4y = 2$

Step 3: Add the two equations.

$-6x + 4y = 2$
$+\,(x - 4y = -7)$ Add equation 2 to equation 1.
$-5x + 0 = -5$ Simplify.
$-5x = -5$ Divide both sides by -5.
$x = 1$

Step 4: Substitute the solution for x back into an equation and solve for y.

Equation 1: $3x - 2y = -1$ Substitute 1 for x.
$3\,(1) - 2y = -1$ Simplify.
$3 - 2y = -1$ Subtract 3 from both sides.
$3 - 3 - 2y = -1 - 3$ Simplify.
$-2y = -4$ Divide both sides by -2.
$y = 2$

Step 5: The solution set is $(1, 2)$. To check, substitute the solution into both of the original equations.

Equation 1:
$3x - 2y = -1$
$3\,(1) - 2\,(2) = -1$
$3 - 4 = -1$
$-1 = -1$

Equation 2:
$-4y = -x - 7$
$-4\,(2) = -1 - 7$
$-8 = -8$

The point $(1, 2)$ is the point of intersection.

For each of the following pairs of equations, find the point of intersection by adding the two equations together.

1. $x + 2y = 8$
   $-x - 3y = 2$
2. $x - y = 5$
   $2x + y = 1$
3. $x - y = -1$
   $x + y = 9$
4. $3x - y = -1$
   $x + y = 13$
5. $-x + 4y = 2$
   $x + y = 8$
6. $x + 4y = 10$
   $x + 7y = 16$
7. $2x - y = 2$
   $4x - 9y = -3$
8. $x + 3y = 13$
   $5x - y = 1$
9. $-x = y - 1$
   $x = y - 1$
10. $x - y = 2$
    $2y + x = 5$
11. $5x + 2y = 1$
    $4x + 8y = 20$
12. $3x - 2y = 14$
    $x - y = 6$
13. $2x + 3y = 3$
    $3x + 5y = 5$
14. $x - 4y = 6$
    $-x - y = -1$
15. $x = 2y + 3$
    $y = 3 - x$

12.4 Graphing Systems of Inequalities

We solve systems of inequalities best graphically. Look at the following example.

Example 6: Sketch the solution set of the following system of inequalities:

$y > -2x - 1$ and $y \leq 3x$

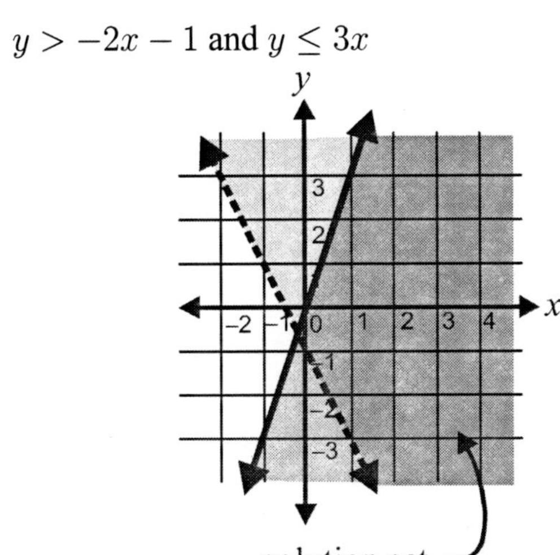

Step 1: Graph both inequalities on a Cartesian plane. Study chapter 10 if you need to review how to graph inequalities.

Step 2: Shade the portion of the graph that represents the solution set to each inequality just as you did in chapter 10.

Step 3: Any shaded region that overlaps is the solution set of both inequalities.

Graph the following systems of inequalities on your own graph paper. Shade and identify the solution set for both inequalities.

1. $2x + 2y \geq -4$
   $3y < 2x + 6$
2. $7x + 7y \leq 21$
   $8x < 6y - 24$
3. $9x + 12y < 36$
   $34x - 17y > 34$
4. $-11x - 22y \geq 44$
   $-4x + 2y \leq 8$
5. $24x < 72 + 36y$
   $11x + 22y \leq -33$
6. $15x - 60 < 30y$
   $20x + 10y < 40$
7. $-12x + 24y > -24$
   $10x < -5y + 15$
8. $y \geq 2x + 2$
   $y < -x - 3$
9. $3x + 4y \geq 12$
   $y > -3x + 2$
10. $-3x \leq 6 + 2y$
    $y \geq -x - 2$
11. $2x - 2y \leq 4$
    $3x + 3y \leq -9$
12. $-x \geq -2y - 2$
    $-2x - 2y > 4$

Chapter 12 Review

For each pair of equations below, tell whether the graphs of the lines will be collinear, parallel, or intersecting.

1. $y = 4x + 1$
   $y = 4x - 3$
2. $y - 4 = x$
   $2x + 8 = 2y$
3. $x + y = 5$
   $x - y = -1$
4. $2y - 3x = 6$
   $4y = 6x + 8$
5. $5y = 3x - 7$
   $4x - 3y = -7$
6. $2x - 2y = 2$
   $y - x = -1$

Find the common solution for each of the following pairs of equations, using the substitution method.

7. $x - y = 2$
   $x + 4y = -3$
8. $x + y = 1$
   $x + 3y = 1$
9. $-4y = -2x + 4$
   $x = -2y - 2$
10. $2x + 8y = 20$
    $5y = 12 - x$
11. $x = y - 3$
    $-x = y + 3$
12. $-2x + y = -3$
    $x - y = 9$

Graph the following systems of inequalities on your own graph paper. Identify the solution set to both inequalities.

13. $x + 2y \geq 2$
    $2x - y \leq 4$
14. $20x + 10y \leq 40$
    $3x + 2y \geq 6$
15. $6x + 8y \leq -24$
    $-4x + 8y \geq 16$
16. $14x - 7y \geq -28$
    $3x + 4y \leq -12$
17. $2y \geq 6x + 6$
    $2x - 4y \geq -4$
18. $9x - 6y \geq 18$
    $3y \geq 6x - 12$

Find the point of intersection for each pair of equations by adding and/or subtracting the two equations.

19. $2x + y = 4$
    $3x - y = 6$
20. $x + 2y = 3$
    $x + 5y = 0$
21. $x + y = 1$
    $y = x + 7$
22. $2x + 4y = 5$
    $3x + 8y = 9$
23. $2x - 2y = 7$
    $3x - 5y = \frac{5}{2}$
24. $x - 3y = -2$
    $y = -\frac{1}{3}x + 4$

Chapter 13
Relations and Functions

This chapter covers the following North Carolina mathematics standards for Algebra I:

Competency Goal	Objectives
Number and Operations	1.01 1.02
Algebra	4.01a 4.04

13.1 Relations

A **relation** is a set of ordered pairs. We call the set of the first members of each ordered pair the **domain** of the relation. We call the set of the second members of each ordered pair the **range**.

Example 1: State the domain and range of the following relation:
$\{(2,4), (3,7), (4,9), (6,11)\}$

Solution: Domain: $\{2,3,4,6\}$ the first member of each ordered pair
Range: $\{4,7,9,11\}$ the second member of each ordered pair

State the domain and range for each relation.

1. $\{(2,5), (9,12), (3,8), (6,7)\}$
2. $\{(12,4), (3,4), (7,12), (26,19)\}$
3. $\{(4,3), (7,14), (16,34), (5,11)\}$
4. $\{(2,45), (33,43), (98,9), (43,61), (67,54)\}$
5. $\{(78,14), (29,67), (84,49), (16,18), (98,46)\}$
6. $\{(-8,16), (23,-7), (-4,-9), (16,-8), (-3,6)\}$
7. $\{(-7,-4), (-3,16), (-4,17), (-6,-8), (-8,12)\}$
8. $\{(-1,-2), (3,6), (-7,14), (-2,8), (-6,2)\}$
9. $\{(0,9), (-8,5), (3,12), (-8,-3), (7,18)\}$
10. $\{(58,14), (44,97), (74,32), (6,18), (63,44)\}$
11. $\{(-7,0), (-8,10), (-3,11), (-7,-32), (-2,57)\}$
12. $\{(18,34), (22,64), (94,36), (11,18), (91,45)\}$

When given an equation in two variables, the domain is the set of x values that satisfies the equation. The range is the set of y values that satisfies the equation.

Example 2: Find the range of the relation $3x = y + 2$ for the domain $\{-1, 0, 1, 2, 3\}$. Solve the equation for each value of x given. The result, the y values, will be the range.

Given:

x	y
-1	
0	
1	
2	
3	

Solution:

x	y
-1	-5
0	-2
1	1
2	4
3	7

The range is $\{-5, -2, 1, 4, 7\}$.

Find the range of each relation for the given domain.

	Relation	Domain	Range
1.	$y = 5x$	$\{1, 2, 3, 4\}$	
2.	$y = \|x\|$	$\{-3, -2, -1, 0, 1\}$	
3.	$y = 3x + 2$	$\{0, 1, 3, 4\}$	
4.	$y = -\|x\|$	$\{-2 - 1, 0, 1, 2\}$	
5.	$y = -2x + 1$	$\{0, 1, 3, 4\}$	
6.	$y = 10x - 2$	$\{-2, -1, 0, 1, 2\}$	
7.	$y = 3\|x\| + 1$	$\{-2, -1, 0, 1, 2\}$	
8.	$y - x = 0$	$\{1, 2, 3, 4\}$	
9.	$y - 2x = 0$	$\{1, 2, 3, 4\}$	
10.	$y = 3x - 1$	$\{0, 1, 3, 4\}$	
11.	$y = 4x + 2$	$\{0, 1, 3, 4\}$	
12.	$y = 2\|x\| - 1$	$\{-2 - 1, 0, 1, 2\}$	

13.2 Determining Domain and Range from Graphs

The domain is all of the x values that lie on the function in the graph from the lowest x value to the highest x value. The range is all of the y values that lie on the function in the graph from the lowest y to the highest y.

Example 3: Find the domain and range of the graph.

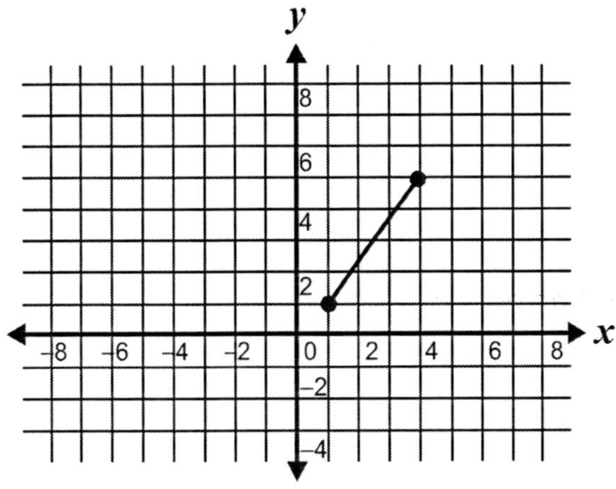

Step 1: First find the lowest x value depicted on the graph. In this case it is 1. Then find the highest x value depicted on the graph. The highest value of x on the graph is 4. The domain must contain all of the values between the lowest x value and the highest x value. The easiest way to write this is $1 \leq \text{Domain} \leq 4$ or $1 \leq x \leq 4$.

Step 2: Perform the same process for the range, but this time look at the lowest and highest y values. The answer is $1 \leq \text{Range} \leq 5$ or $1 \leq y \leq 5$.

Find the domain and range of each graph below. Write your answers in the line provided.

1.

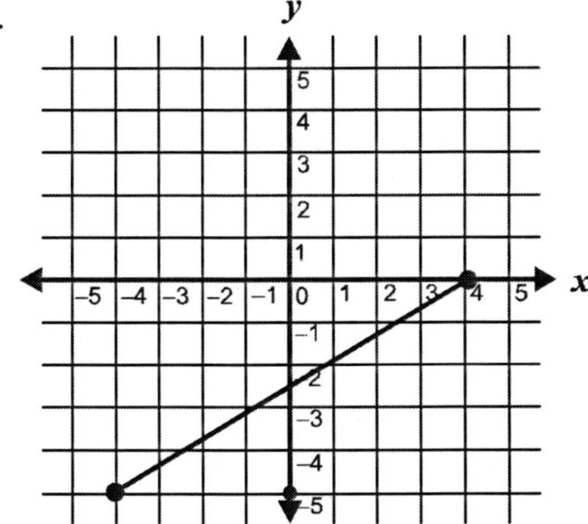

____

2.

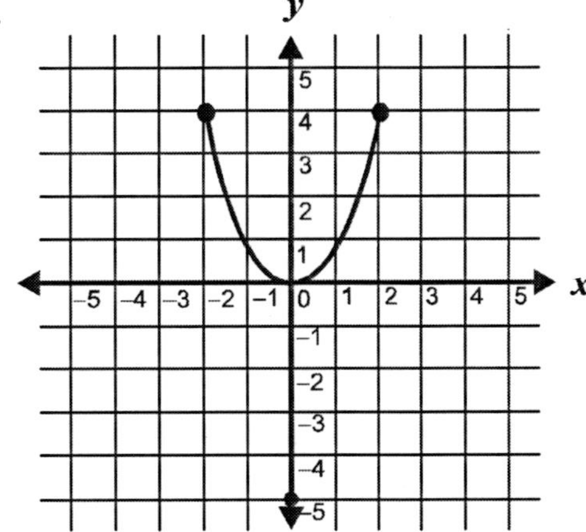

____

3.

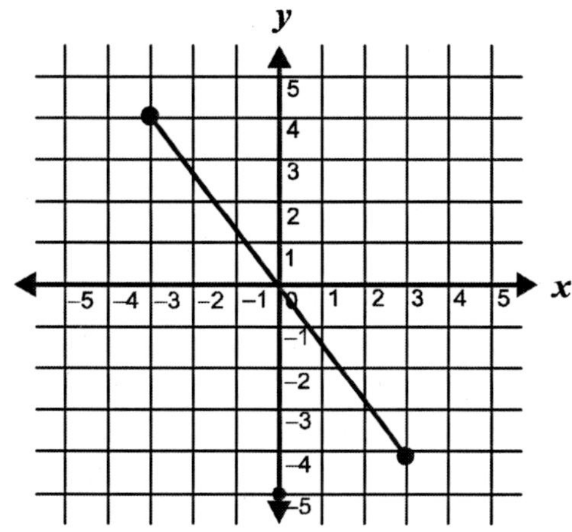

6.

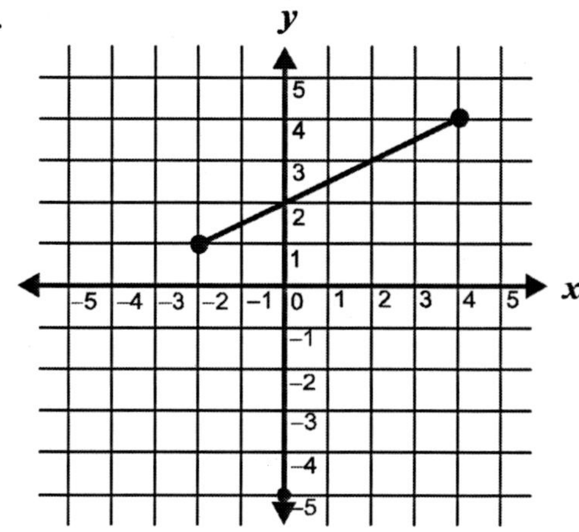

4.

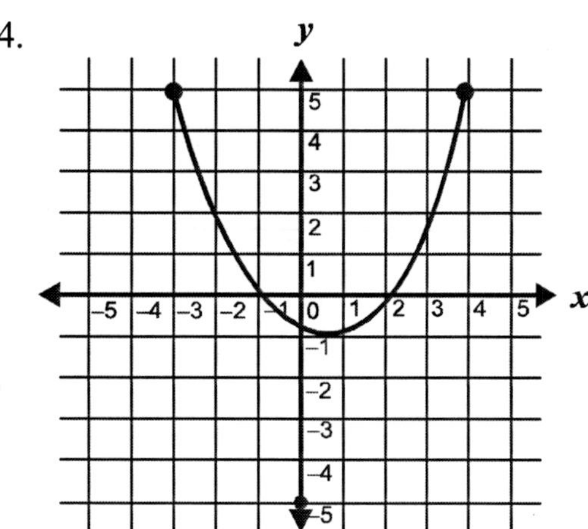

7.

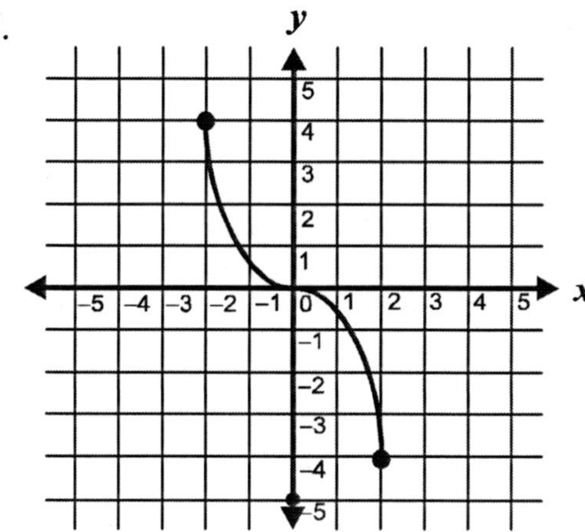

5.

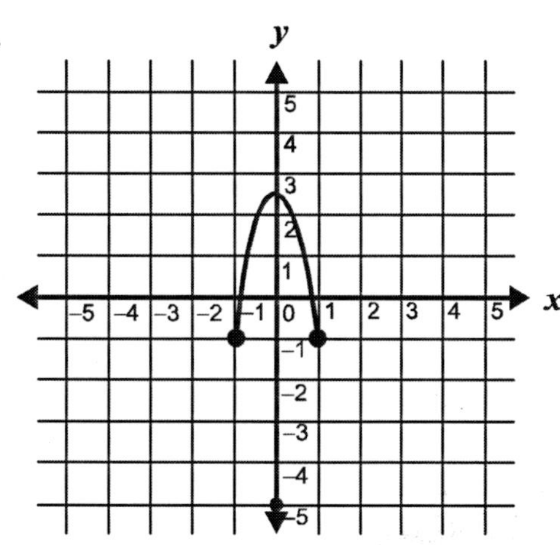

8.

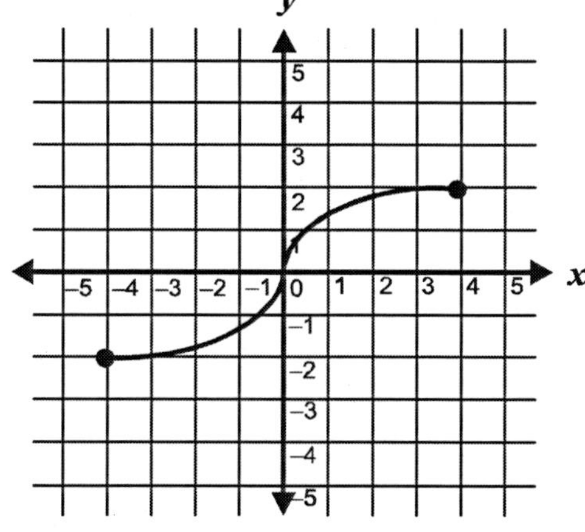

13.3 Functions

Some relations are also **functions**. A relation is a function if **for every element in the domain, there is exactly one element in the range**. In other words, for each value for x there is only one unique value for y.

Example 4: $\{(2,4),(2,5),(3,4)\}$ is **NOT** a function because in the first pair, 2 is paired with 4, and in the second pair, 2 is paired with 5. The 2 can be paired with only one number to be a function. In this example, the x value of 2 has more than one value for y: 4 and 5.

Example 5: $\{(1,2),(3,2),(5,6)\}$ **IS** a function. Each first number is paired with only one second number. The 2 is repeated as a second number, but the relation remains a function.

Determine whether the ordered pairs of numbers below represent a function. Write "F" if it is a function. Write "NF" if it is not a function.

1. $\{(-1,1),(-3,3),(0,0),(2,2)\}$ ______
2. $\{(-4,-3),(-2,-3),(-1,-3),(2,-3)\}$ ______
3. $\{(5,-1),(2,0),(2,2),(5,3)\}$ ______
4. $\{(-3,3),(0,2),(1,1),(2,0)\}$ ______
5. $\{(-2,-5),(-2,-1),(-2,1),(-2,3)\}$ ______
6. $\{(0,2),(1,1),(2,2),(4,3)\}$ ______
7. $\{(4,2),(3,3),(2,2),(0,3)\}$ ______
8. $\{(-1,-1),(-2,-2),(3,-1),(3,2)\}$ ______
9. $\{(2,-2),(0,-2),(-2,0),(1,-3)\}$ ______
10. $\{(2,1),(3,2),(4,3),(5,-1)\}$ ______
11. $\{(-1,0),(2,1),(2,4),(-2,2)\}$ ______
12. $\{(1,4),(2,3),(0,2),(0,4)\}$ ______
13. $\{(0,0),(1,0),(2,0),(3,0)\}$ ______
14. $\{(-5,-1),(-3,-2),(-4,-9),(-7,-3)\}$ ______
15. $\{(8,-3),(-4,4),(8,0),(6,2)\}$ ______
16. $\{(7,-1),(4,3),(8,2),(2,8)\}$ ______
17. $\{(4,-3),(2,0),(5,3),(4,1)\}$ ______
18. $\{(2,-6),(7,3),(-3,4),(2,-3)\}$ ______
19. $\{(1,1),(3,-2),(4,16),(1,-5)\}$ ______
20. $\{(5,7),(3,8),(5,3),(6,9)\}$ ______

13.4 Function Notation

Function notation is used to represent relations which are functions. Some commonly used letters to represent functions include f, g, h, F, G, and H.

Example 6: $f(x) = 2x - 1$; find $f(-3)$

Step 1: Find $f(-3)$ means to replace x with -3 in the relation $2x - 1$.
$f(-3) = 2(-3) - 1$

Step 2: Solve $f(-3)$. $f(-3) = 2(-3) - 1 = -6 - 1 = -7$
$f(-3) = -7$

Example 7: $g(x) = 4 - 2x^2$; find $g(2)$

Step 1: Replace x with 2 in the relation $4 - 2x^2$.
$g(2) = 4 - 2(2)^2$

Step 2: Solve $g(2)$. $g(2) = 4 - 2(2)^2 = 4 - 2(4) = 4 - 8 = -4$
$g(2) = -4$

Find the solutions for each of the following.

1. $F(x) = 2 + 3x^2$; find $F(3)$
2. $f(x) = 4x + 6$; find $f(-4)$
3. $H(x) = 6 - 2x^2$; find $H(-1)$
4. $g(x) = -3x + 7$; find $g(-3)$
5. $f(x) = -5 + 4x$; find $F(7)$
6. $G(x) = 4x^2 + 4$; find $G(0)$
7. $f(x) = 7 - 6x$; find $f(-4)$
8. $h(x) = 2x^2 + 10$; find $h(5)$
9. $F(x) = 7 - 5x$; find $F(2)$
10. $f(x) = -4x^2 + 5$; find $f(-2)$

13.5 Recognizing Functions

Recall that a relation is a function with only one y value for every x value. We can depict functions in many ways including through graphs.

Example 8:

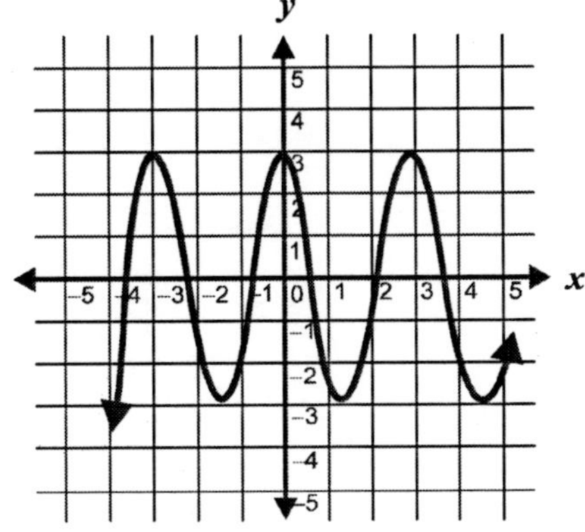

This graph **IS** a function because it has only one y value for each value of x.

Example 9:

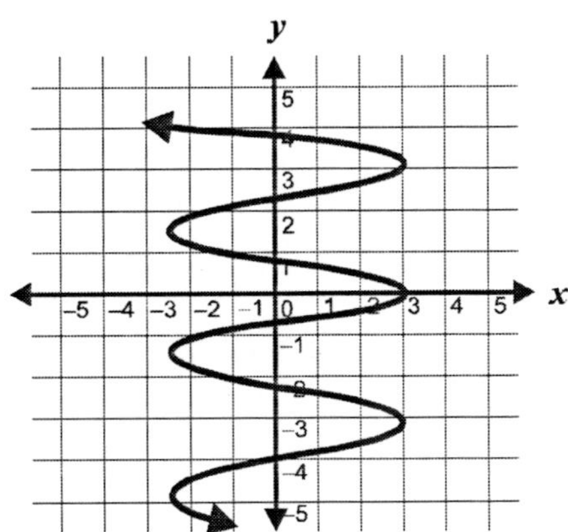

This graph is **NOT** a function because there is more than one y value for each value of x.

Hint: An easy way to determine a function from a graph is to do a vertical line test. First, draw a vertical line that crosses over the whole graph. If the line crosses the graph more than one time, then it is not a function. If it only crosses it once, it is a function. Take Example 9 above:

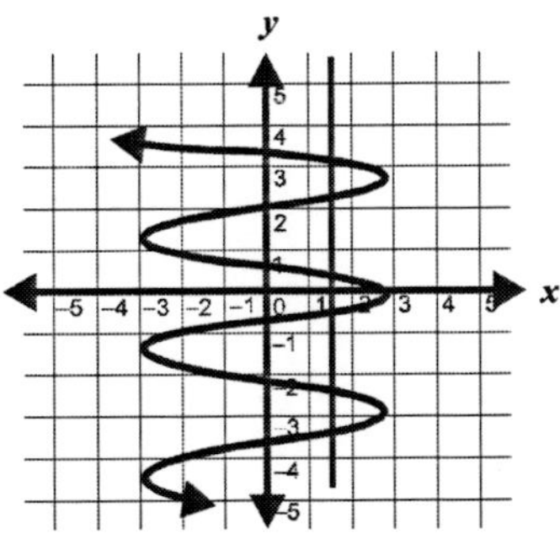

Since the vertical line passes over the graph six times, it is not a function.

Determine whether or not each of the following graphs is a function. If it is, write function on the line provided. If it is not a function, write NOT a function on the line provided.

1.

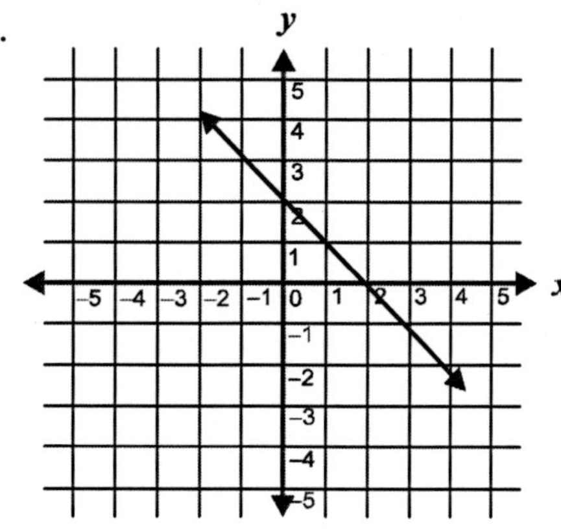

____

4.

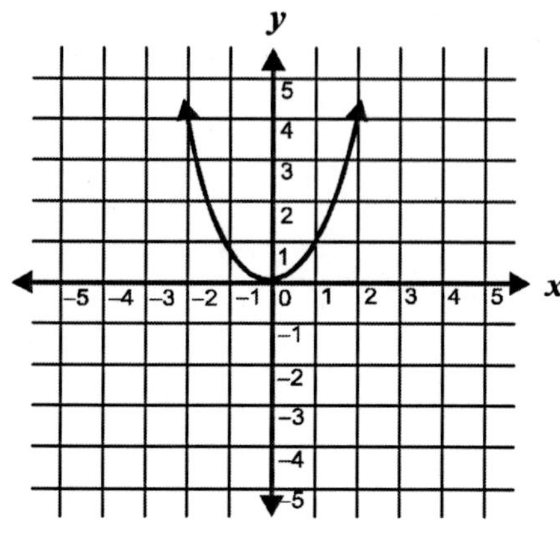

____

2.

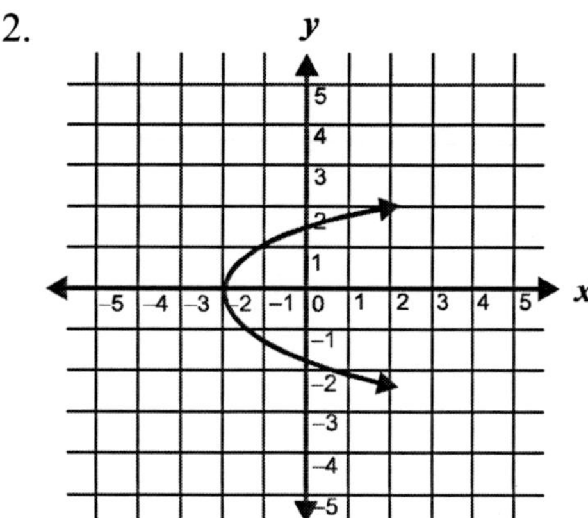

____

5.

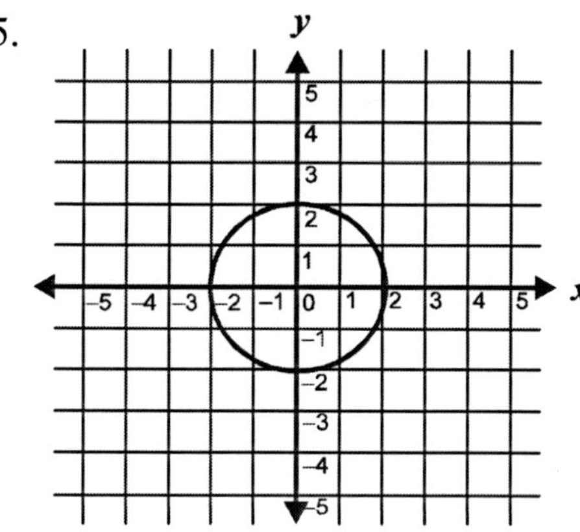

____

3.

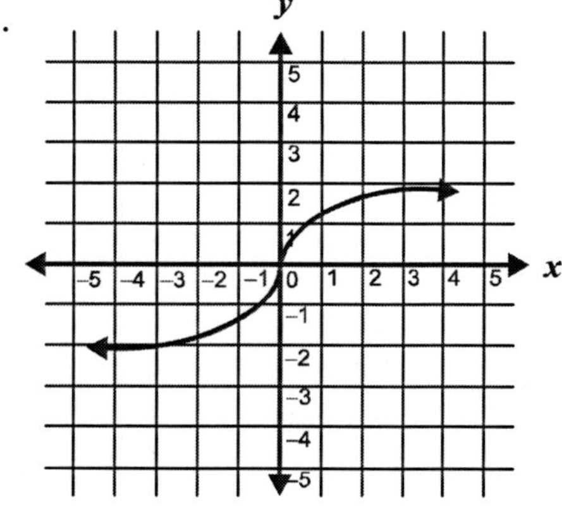

____

6.

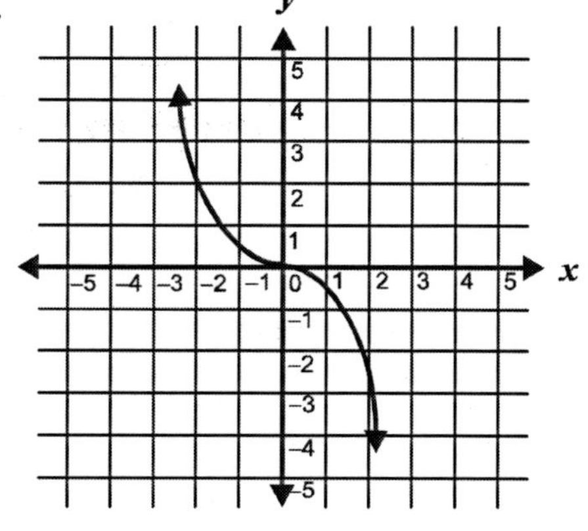

____

7.

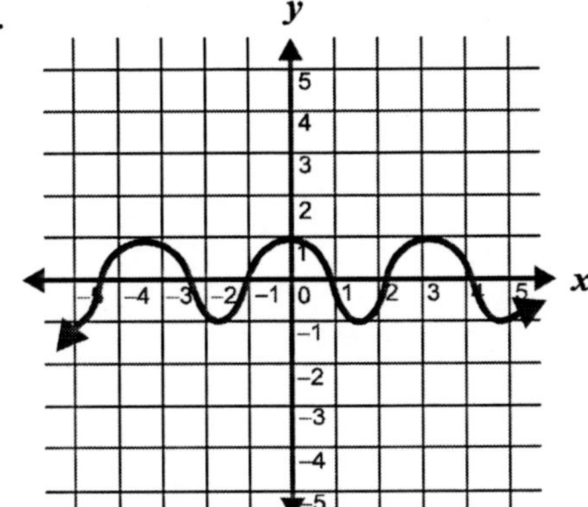

10.

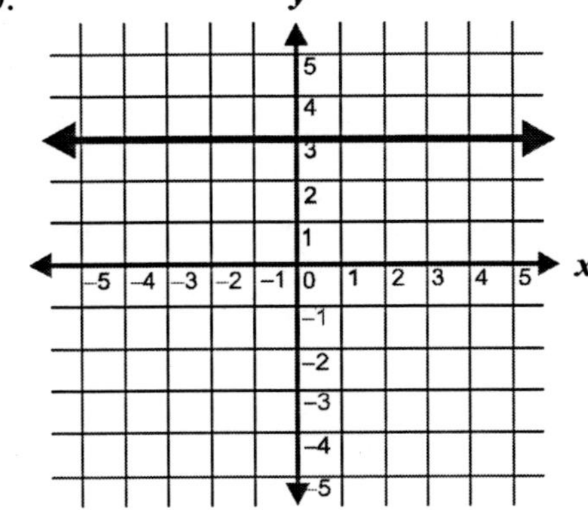

8.

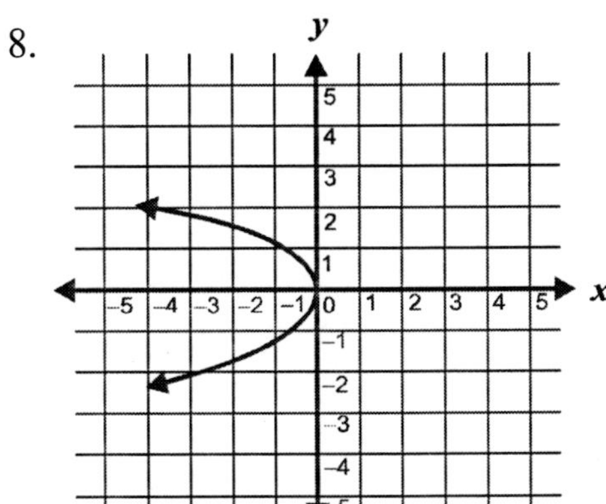

11.

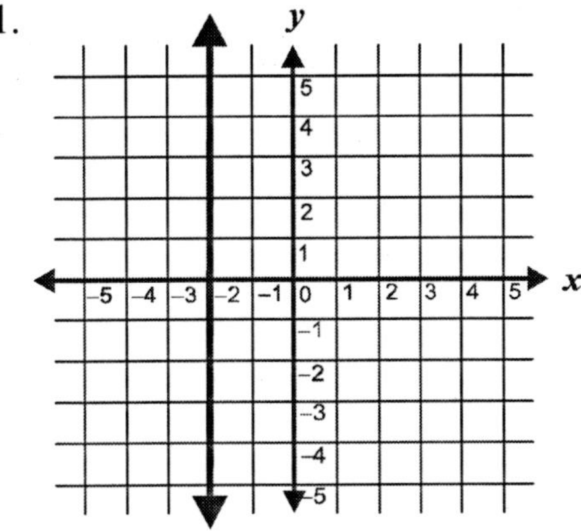

9.

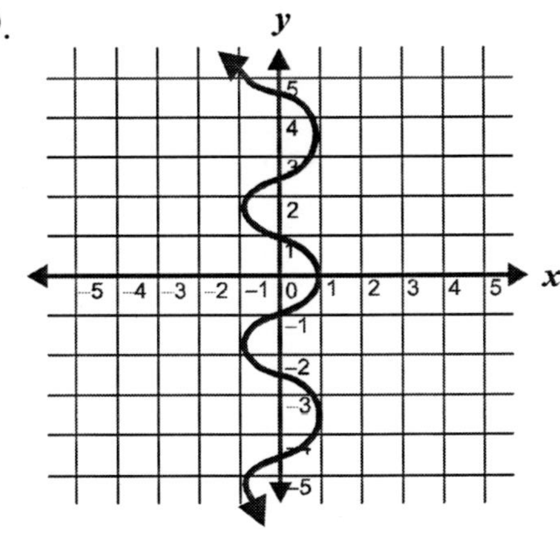

12.

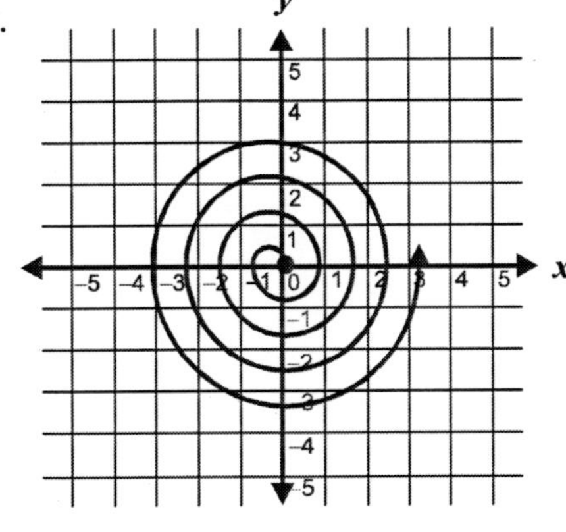

13.6 Independent and Dependent Variables

As stated previously, a relation is a function if for every element in the domain there is exactly one element in the range. The domain values are generally known, and the range values are determined by solving the function. As each domain value is applied to the function, only one range value will result. The variable that is used to represent the domain values is called the **independent variable** because it is not dependent on any other value. The variable that is used to represent the range values is called the **dependent variable** because its value will be determined by its corresponding domain value.

Example 10: Mrs. Alexander assigned to her students an open book quiz containing 35 questions to be completed at home. Those students who returned the completed quiz by the due date would receive 30 points for turning the assignment in on time and 2 points for each correct answer. A student's grade on the open book quiz can be expressed as the function $f(a) = 30 + 2a$, where a represents the number of correct answers. Identify the independent and dependent variables in this function.

Solution: The independent variable in this problem is a, the number of correct answers because it is not dependent on any other value in the function.
The dependent variable in this problem is the grade, $f(a)$, because it is dependent on the number of correct answers. The dependent variable could have also been assigned a variable such as G, T, or y. Using function notation clearly illustrates in the algebraic sentence that the dependent variable is a function of the independent variable.

Identify the independent and dependent variables in the following functions.

1. A local bookstore is encouraging its customers to drop off used books to be given to schools, libraries, and other community organizations. They are offering to anyone who drops off books a special hard-cover edition of *Oliver Twist* for \$25.95 minus \$0.10 for each used book. The cost for the special edition of *Oliver Twist* can be expressed as $G(u) = \$25.95 - \$0.10u$.

2. Claudia is planning a surprise birthday party for her best friend. To make sure that she has enough food, she is ordering 1 sub sandwich for every person who is coming to the party plus an additional 10 sub sandwiches. The number of sandwiches Claudia is ordering can be written algebraically as follows: $s = 10 + p$.

3. John and Mike are brothers who are training for their school swim team. John has been swimming longer than Mike and is able to swim more laps. For every lap that Mike swims, John swims 3, and the number of laps that John swims can be expressed as $j = 3m$.

Write a function for each of the following word problems. Identify the independent and dependent variables.

4. All delivery drivers at Victor's Pizza Pub are hired to work 5-hour shifts. For each shift worked, a delivery driver gets paid $40 plus $2 for every pizza delivered. Write a function that expresses a delivery driver's earnings for one shift.

5. At 8:00 AM the temperature outside was 50°. As the morning progressed, the temperature rose by 3° every hour. Write a function that describes the temperature at any given hour after 8:00 AM.

6. Austin wanted to borrow $325 from his father to buy a new mountain bike. His father agreed, if Austin would pay off the debt by doing odd jobs around the house and in the yard for a wage of $6.50/hour. Write a function that will help Austin calculate how much debt he has left to pay his father.

7. A new shopping center is leasing store space at a monthly rate of $3.00/ft^2. Each individual store will be 20 ft wide, but the length will vary. Write a function that expresses the monthly lease rate of any individual store. Remember that Area = Length × Width.

8. Every year a professional baseball player gives $10,000 to a national research fund. He also gives $1,500 for every home run he hits. Express the baseball player's contributions as an algebraic sentence.

9. The local natural gas company charges a monthly usage fee of $25.00. In addition, each household is charged $.67 per therm of natural gas used during the month. Write a function that a homeowner could use to calculate his/her monthly gas bill.

10. Boy Scout Troop 575 is planning an exciting summer mountain adventure. To raise money for the trip, the boys are selling popcorn. Each member of the troop must pay $400 for the trip. For each case of popcorn a Boy Scout sells, $10 will be applied toward his trip fees. Write a formula that describes the amount of money a Boy Scout must pay out of pocket.

11. Oak Hills High School is putting on a spring musical. Tickets are being sold for $6.50 per person. The drama club at Oak Hills gets $\frac{1}{3}$ of the total ticket sales to use for future programs. Write a function that expresses how much money the drama club will receive from the spring musical.

12. Josie rented a car for one day from a company that charges $30 per day plus $.20 per mile. What function would Josie use to calculate her total bill before taxes?

13. Hannah wanted to participate in a yard sale being sponsored by her school. She would have to pay $5.00 to rent the space for her items and then would receive 65% of the money her items generated. The remaining 35% would be given to a local charity. Write a function that expresses Hannah's net profit.

13.7 Relations That Can Be Represented by Functions

Real-life examples can be represented by functions. The most common functions are exponential growth and decay and half-life.

Example 11: Atlanta, GA has a population of about 410,000 people. The U.S. Census Bureau estimates that the population will double in 26 years. If the population continues at the same rate, what will the population be in
a) 10 years?
b) 50 years?

Step 1: Use the double growth equation $P = P_0(2^{t/d})$, where P = population at time t, P_0 = population at time $t = 0$, and d = double time.

Step 2: Determine the variable of each of the facts given in the problem. In this case, $P_0 = 410,000$ people, $d = 26$ years, and $t = 10$ years for part a and $t = 50$ years for part b.

Step 3: Plug all of the information into the given equation. Round to the nearest whole number.
a) $P = 410,000(2^{10/26}) = 410,000\,(1.3055) = 535,260$ people
b) $P = 410,000(2^{50/26}) = 410,000\,(3.7923) = 1,554,847$ people

Find the answers to the real-life problems by using the equations and variables given. Round your answer to the nearest whole numbers.

For questions 1 and 2 use the following half-life formula.

$A = A_0\left(\frac{1}{2}\right)^{t/h}$
A = amount at time t
A_0 = amount at time $t = 0$
h is the half-life

1. If you have 6,000 atoms of hydrogen (H), and hydrogen's half-life is 12.3 years, how many atoms will you have left after 7 years?

2. Chlorine (Cl) has a half-life of 55.5 minutes. If you start with 200 milligrams of chlorine, how many will be left after 5 hours?

For questions 3 and 4 use the double growth formula.

$P = P_0(2)^{t/d}$
$P =$ amount at time t
$P_0 =$ amount at time $t = 0$
d is the half-life

3. There are about $3,390,000$ Girl Scouts in the United States. The Girl Scout Council says that there is a growth rate of $5 - 10\%$ per year, so they expect the Girl Scout population in the United States to double in 12 years. If the Girl Scout's organization expands as continuously as it has been, what will the population be
   (A) in 8 years?
   (B) next year?

4. Dr. Kellie noticed the bacteria growth in her laboratory. After observing the bacteria, she concluded that the double time of the bacteria is 40 minutes, and she started off with just $2,500$ bacteria. Assuming this information is accurate and constant, how many bacteria will be in Dr. Kellie's lab
   (A) in 5 minutes?
   (B) after 3 hours?

For questions 5 and 6 use the compound interest formula.

$A = P\left(1 + \frac{r}{k}\right)^{kt}$
$A =$ amount at time, t
$P =$ principle amount invested
$k =$ how many times per year interest is compounded
$r =$ rate

5. Lisa invested $\$1,000$ into an account that pays 6% interest compounded monthly. If this account is for her newborn, how much will the account be worth on his 21st birthday, which is exactly 21 years from now?

6. Mr. Dumple wants to open up a savings account. He has looked at two different banks. Bank 1 is offering a rate of 5% compounded daily. Bank 2 is offering an account that has a rate of 8%, but is only compounded semi-yearly. Mr. Dumple puts $\$5,000$ in an account and wants to take it out for his retirement in 10 years. Which bank will give him the most money back?

13.8 Exponential Growth and Decay

Many quantities experience exponential growth or decay under certain conditions. Examples include bacteria, populations, disease, money in a savings account that compounds interest, and radioisotopes. Exponential functions are those functions in which the independent variable is time, and time is an exponent (thus the name exponential function). For instance, the formula for growth of money in a savings account that compounds interest annually is:

$$A = P(1 + r)^t$$

where A is the value of the account after t years, P is the original amount of money in the account, and r is the annual interest rate.

Below are graphs of the general forms of exponential growth functions and exponential decay functions. Time is represented on the x-axis. Whatever is growing or decaying exponentially, such as population or money, is represented on the y-axis. Note that exponential function graphs are generally in Quadrant I since time and objects cannot be assigned negative values.

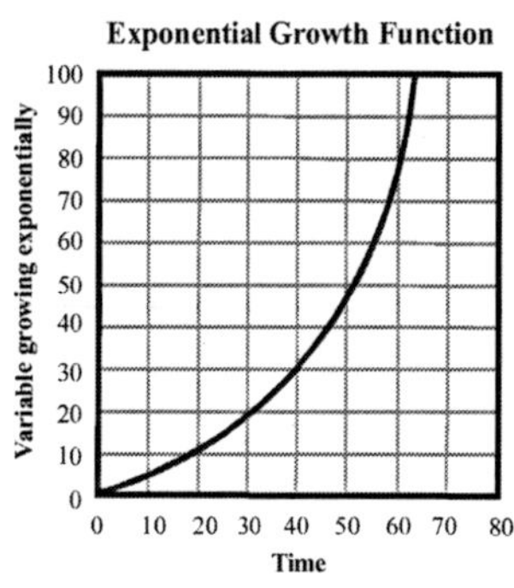

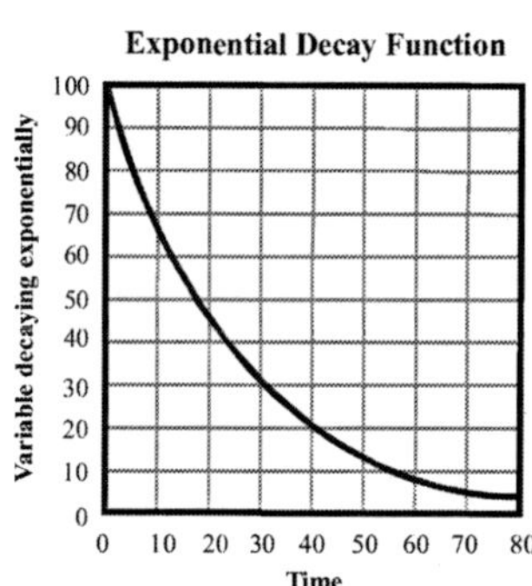

Example 12: Mason deposited $\$2,000$ into a savings account that pays an annual interest rate of 9% compounded annually. Using the formula $A = P(1 + r)^t$ determine the amount of money in the savings account after 1 year, 5 years, and 20 years. Using the calculated values, construct a graph.

Step 1: Consider the known values. $P = 2,000$, and $r = 0.09$. The problem will have to be worked three times where $t = 1$, $t = 5$, and $t = 20$. A is the amount being calculated.

Step 2:

$A = 2000(1 + 0.09)^1$	$A = 2000(1 + 0.09)^5$	$A = 2000(1 + 0.09)^{20}$
$A = 2000(1.09)^1$	$A = 2000(1.09)^5$	$A = 2000(1.09)^{20}$
$A = \$2,180$	$A = \$3,077.25$	$A = \$11,208.82$

Step 3: Use the calculated values to graph the function.

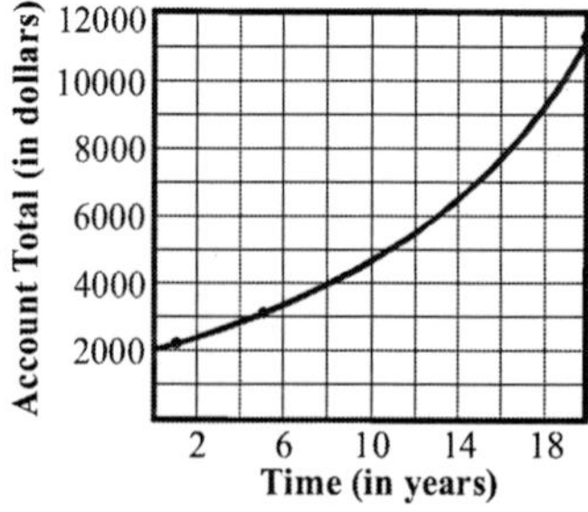

Fill in the tables for the following functions. On the line under each table, label the given function as an exponential growth function or an exponential decay function. Round your answers to two decimal places. For extra practice, graph the functions.

1. $F(t) = 15(1.01)^t$

t	$F(t)$
1	
2	
3	
4	

____

3. $M(t) = 1000(1.04)^t$

t	$M(t)$
2	
4	
6	
8	

____

5. $C(t) = 5300(0.5)^t$

t	$C(t)$
5	
10	
15	
20	

____

2. $S(t) = 350(0.85)^t$

t	$S(t)$
1	
3	
5	
7	

____

4. $B(t) = 2(2.50)^t$

t	$B(t)$
1	
2	
3	
4	

____

6. $R(t) = 80\left(\frac{1}{3}\right)^t$

t	$R(t)$
2	
4	
6	
8	

____

Refer to the graph at right to answer question 7–10.

7. Which town is experiencing exponential decay? growth?

8. Considering both towns A and B, what is changing exponentially with time?

9. Why would it not make sense to draw the graph of town B below the x-axis?

10. In what year does the population of town B reach 3,000?

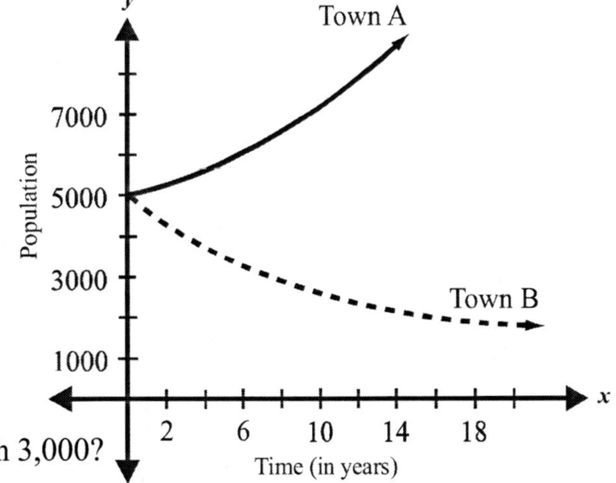

Chapter 13 Review

1. What is the domain of the following relation? $\{(-1,2),(2,5),(4,9),(6,11)\}$
2. What is the range of the following relation? $\{(0,-2),(-1,-4),(-2,6),(-3,-8)\}$
3. Find the range of the relation $y = 5x$ for the domain $\{0,1,2,3,4\}$.
4. Find the values of $M(y)$ of the relation $M(y) = 2(1.1)^y$ for the domain $\{2,3,4,5,6\}$.
5. Find the range of the following relation for the domain $\{0,2,6,8,10\}$. $B(t) = 600(0.75)^t$
6. Find the range of the relation $y = \dfrac{3(x-2)}{5}$ for the domain $\{-8,-3,7,12,17\}$.
7. Find the range of the relation $y = 10 - 2x$ for the domain $\{-8,-4,0,4,8\}$.
8. Find the range of the relation $y = \dfrac{4+x}{3}$ for the domain $\{-7,-1,2,5,8\}$.

For each of the following relations given in questions 9–13, write F if it is a function and NF if it is not a function.

9. $\{(1,2),(2,2),(3,2)\}$
10. $\{(-1,0),(0,1),(1,2),(2,3)\}$
11. $\{(2,1),(2,2),(2,3)\}$
12. $\{(1,7),(2,5),(3,6),(2,4)\}$
13. $\{(0,-1),(-1,-2),(-2,-3),(-3,-4)\}$

For questions 14–19, find the range of the following functions for the given value of the domain.

14. For $g(x) = 2x^2 - 4x$; find $g(-1)$
15. For $h(x) = 3x(x-4)$; find $h(3)$
16. For $f(n) = \dfrac{1}{n+3}$; find $f(4)$
17. For $G(n) = \dfrac{2-n}{2}$; find $G(8)$
18. For $H(x) = 2x(x-1)$; find $H(4)$
19. For $f(x) = 7x^2 + 3x - 2$; find $f(2)$

20. Trent sells computers and other electronic devices for Computer Town. He receives \$300 per week and 60% of his total sales. Write a function that expresses Trent's weekly earnings. Identify the independent and dependent variables.

Practice Test 1

1. What is the slope of a line perpendicular to the line passing through the points $(3, 6)$ and $(5, 1)$?

   (A) $-\frac{5}{2}$

   (B) $-\frac{4}{3}$

   (C) $-\frac{3}{4}$

   (D) $\frac{2}{5}$

   4.01b

2. Which of the following is a number which, when squared, results in a number less than itself?

   (A) -4
   (B) 4^{-2}
   (C) 4
   (D) $-\frac{1}{4}$

   1.01a

3. What is the slope of a line parallel to a line having slope $-\frac{3}{2}$?

   (A) -6

   (B) $-\frac{3}{2}$

   (C) $-\frac{2}{3}$

   (D) $\frac{2}{3}$

   2.02

4. The sum of two numbers is fourteen. The sum of six times the smaller number and two equals four less than the product of three and the larger number. Find the two numbers.

   (A) 6 and 8
   (B) 5 and 9
   (C) 3 and 11
   (D) 4 and 10

   4.03

5. Solve: $\dfrac{3x+6}{-2} > -12$

   (A) $x < 24$
   (B) $x > 0$
   (C) $x > 6$
   (D) $x < 6$

   4.01a

6. Simplify: $\dfrac{(3a^2)^3}{a^3}$

   (A) $27a^3$

   (B) $\dfrac{9a^6}{a^3}$

   (C) $9a^3$

   (D) $\dfrac{3a^6}{a^3}$

   1.01a

7. Solve for x: $2(x+5) + 4(2x-1) = -14$

   (A) $x = -2$
   (B) $x = -1$
   (C) $x = -1\frac{4}{5}$
   (D) $x = -1\frac{2}{10}$

   4.01a

8. What is the slope of the equation graphed below?

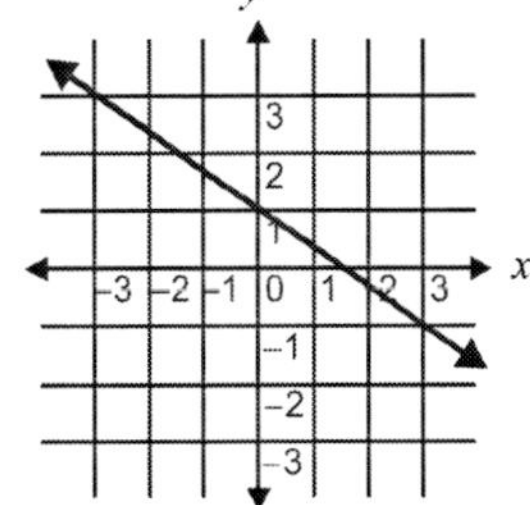

   (A) $\frac{2}{3}$

   (B) $\frac{3}{2}$

   (C) $-\frac{2}{3}$

   (D) $-\frac{3}{2}$

   4.01b

9. Solve: $-6 - x \geq 7$

(A) $x \geq -13$
(B) $x \leq 13$
(C) $x \leq -13$
(D) $x \geq 13$

4.01a

10. There are three brothers. Fernando is two years older than Pedro. Pedro is two years older than Samuel. Together their ages add up to 63 years. How old is Samuel?

(A) 17
(B) 19
(C) 21
(D) 23

4.03

11. Solve $y^2 - 4y - 12 = 0$

(A) $\{2, -6\}$
(B) $\{-2, 6\}$
(C) $\{3, -4\}$
(D) $\{-3, 4\}$

4.02

12. Translate "eighty-four less the product of six and seven" into an algebraic expression.

(A) $(6 \times 7) - 84$
(B) $(6 \times 7)(-84)$
(C) $84 - (6 \times 7)$
(D) $84 \times (-6 + 7)$

1.02

13. Solve: $2(5x - 3) - 6x = 2$

(A) $-\frac{1}{4}$

(B) $\frac{5}{4}$

(C) 1

(D) 2

4.01a

14. The Rockbottom Blues Band charges a \$300 set up fee plus \$175 per hour (h) that they play. Which statement represents the total cost (c) for hiring the band?

(A) $c = 175 + 300h$
(B) $c = (175 + 300)h$
(C) $c = 300 + 175h$
(D) $c = 300 + 175 + h$

4.01b

15. A racetrack timekeeper records the engine horsepower and top speed of three race cars in the table below.

Engine Horsepower	Top Speed (miles per hour)
140	125
160	130
200	140

Which of these graphs correctly represents the top speed as a linear function of the horsepower of the engine?

(A)
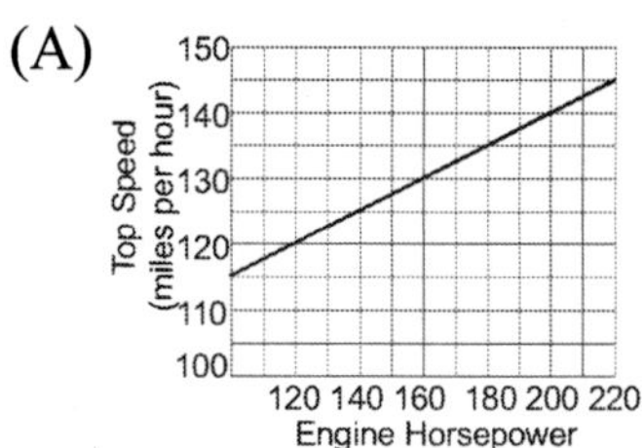

(B)
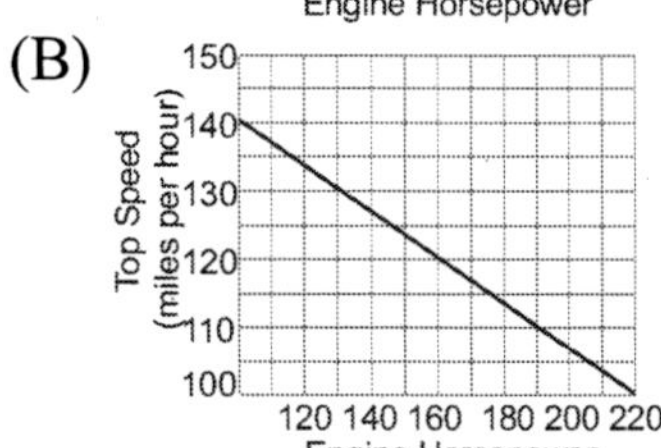

(C)
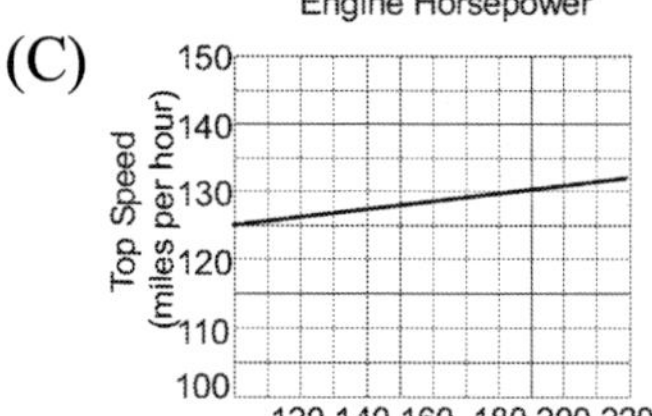

(D)
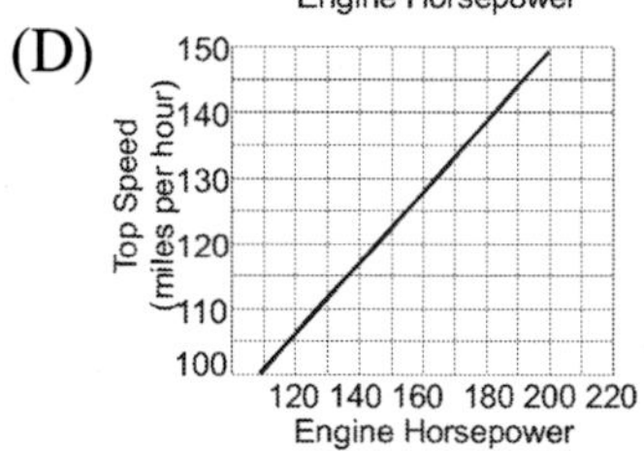

3.03b

16. What transformation of the graph occurs when the graph $y = x - 1$ is changed to $y = 3x - 1$?

(A) The graph shifts down 2 units.
(B) The graph shifts up 2 units.
(C) The slope decreases.
(D) The slope increases.

4.01b

17. Someone reports a fire at the location plotted on the grid.

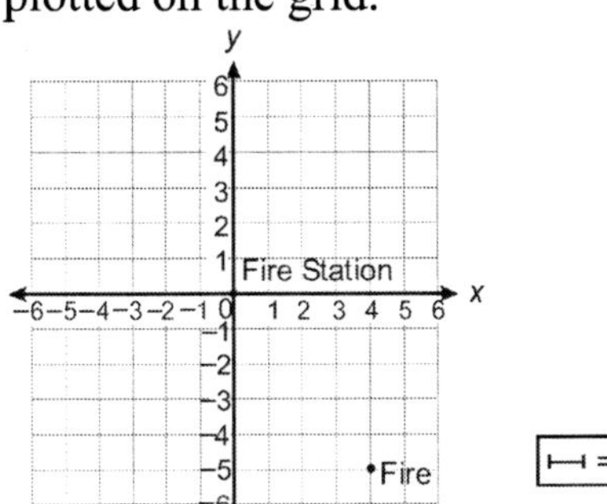

How far is the fire from the fire station?

(A) 3 miles
(B) $\sqrt{20}$ miles
(C) $\sqrt{41}$ miles
(D) 9 miles

2.01

18. If $3x + 4y = 9$, then x equals

(A) $3 - 4y$
(B) $9 - 4y$
(C) $\dfrac{9 + 4y}{3}$
(D) $\dfrac{9 - 4y}{3}$

4.01a

19. If $x = -3$, find $3x^2 - 5x$

(A) 12
(B) -6
(C) 42
(D) 3

4.02

20. Boyle's Law is stated by the formula, $P_1V_1 = P_2V_2$. Find V_1 when $P_1 = 110$, $P_2 = 50$, and $V_2 = 440$.

(A) 110
(B) 200
(C) 220
(D) $21,890$

1.03

21. Solve for a: $-4a - 12 = -36$

(A) 6
(B) -6
(C) 12
(D) -12

4.01a

22. Find $(4y^4 + 2y^2 + 7) + (2y^3 + 5y^2 - 4)$.

(A) $4y^4 + 2y^3 + 7y^2 + 3$
(B) $4y^4 + 4y^3 + 5y^2 + 3$
(C) $8y^7 + 10y^4 - 28$
(D) $8y^{12} + 10y^4 + 3$

1.01b

23. Find $(-3a^2 + 8a - 2) = (-4a^2 - 2a + 6)$.

(A) $a^2 + 10a - 8 = 0$
(B) $-7a^2 + 6a + 4 = 0$
(C) $12a^4 - 16a^2 - 12 = 0$
(D) $a^2 + 6a - 8 = 0$

1.01b

24. Use the correct order of operations to evaluate the following expression.
$4\,(4x - 3)^2$

(A) $16x^2 - 24x + 9$
(B) $400x^2 - 225$
(C) $80x - 45$
(D) $64x^2 - 96x + 36$

1.01c

25. Which of the following is the graph of the equation $y = x + 2$?

(A)

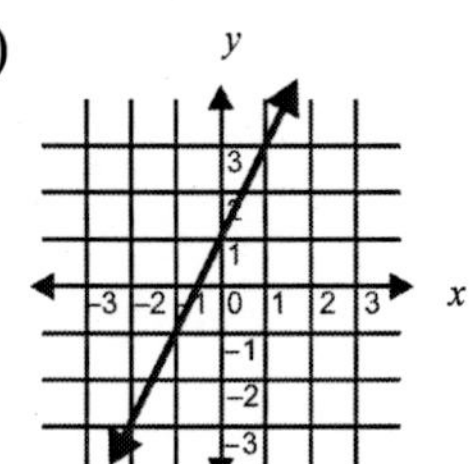

(B)

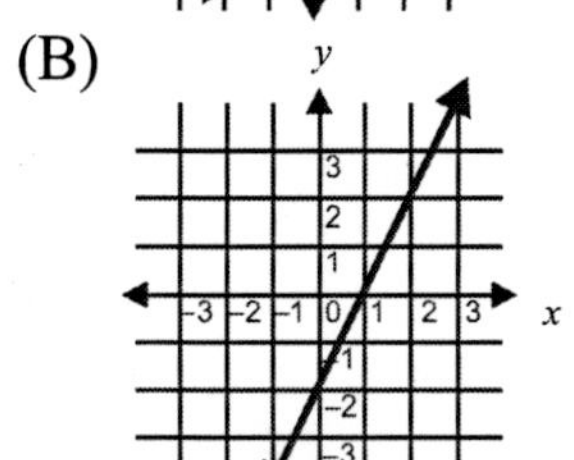

(C)

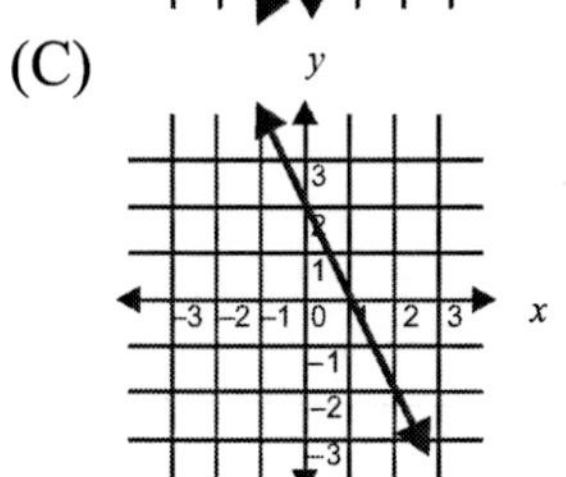

(D)

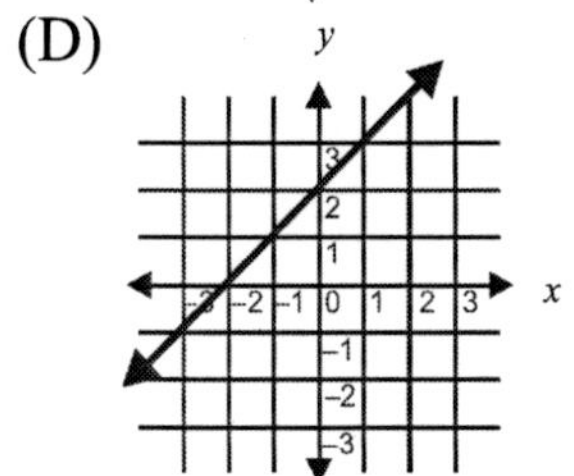

4.01a

26. What is the slope of the equation graphed below ?

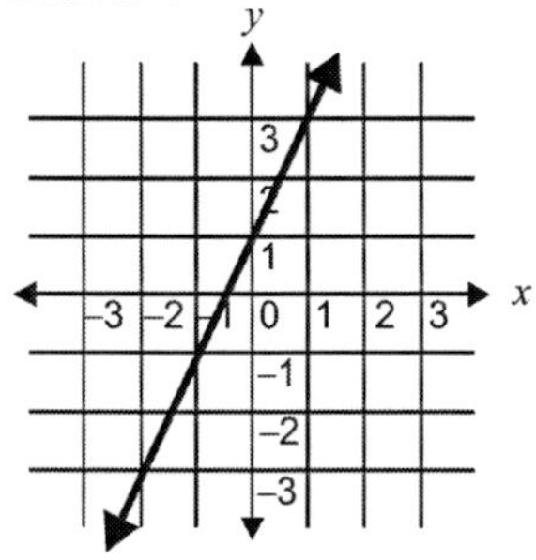

(A) $\frac{1}{2}$

(B) 2

(C) -2

(D) $-\frac{1}{2}$

4.01b

27. Which of the following graphs is not a function?

(A)

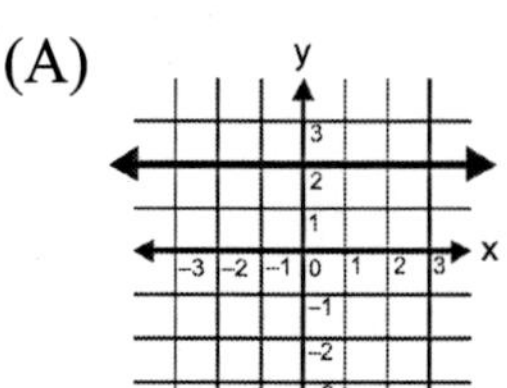

(B)

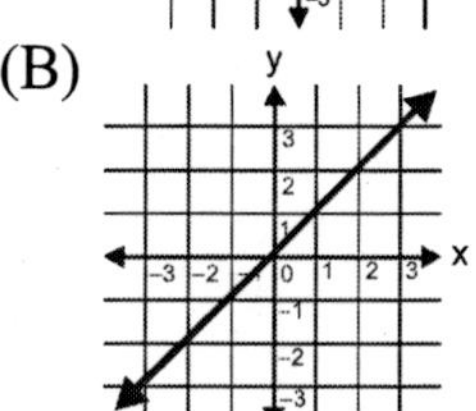

(C)

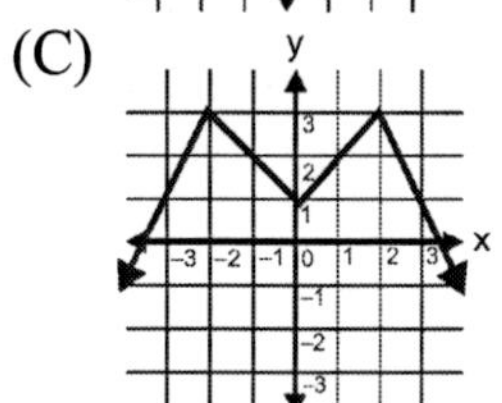

(D)

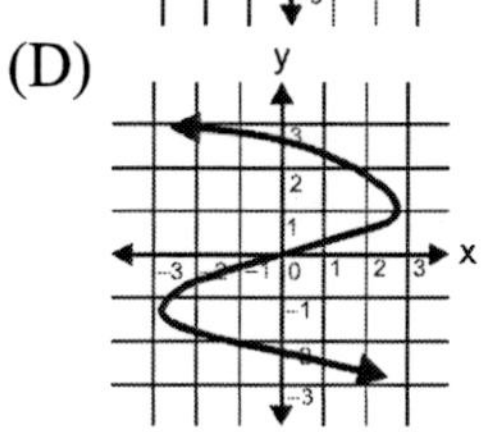

4.01a

28. Solve: $-\dfrac{6}{x} = 12$

(A) 2

(B) -2

(C) $-\frac{1}{2}$

(D) $\frac{1}{2}$

4.01a

29. Solve for x in the following equation.

$$\frac{6x - 40}{2} = 4$$

(A) 6

(B) $\frac{32}{6}$

(C) 8

(D) $7\frac{1}{3}$

4.01a

30. Solve for x: $7(2x+6)-4(9x+6) < -26$

(A) $x > -2$
(B) $x > 2$
(C) $x < -2$
(D) $x < -1$

4.01a

31. Which ordered pair is a solution for the following system of equations?

$-x+7y=4$
$x-5y=-14$

(A) $(3,1)$
(B) $(11,-5)$
(C) $(-18,-2)$
(D) $(-39,-5)$

4.03

32. If the equation below were graphed, which of the following points would lie on the line?

$4x+7y=56$

(A) $(7,4)$
(B) $(0,14)$
(C) $(8,0)$
(D) $(4,7)$

4.01a

33. Multiply and simplify: $(3x+2)(x-4)$

(A) $3x^2-10x-8$
(B) $3x^2+5x-8$
(C) $3x^2+5x-6$
(D) $8x^2-2$

1.01c

34. Solve the equations by substitution.

$-2x-4y=-14$
$5x+y=-1$

(A) $(-1,4)$
(B) $(5,1)$
(C) $(10,-11)$
(D) $(3,2)$

4.03

35. Which of the following is a graph of the inequality $-y \geq 2$?

(A)

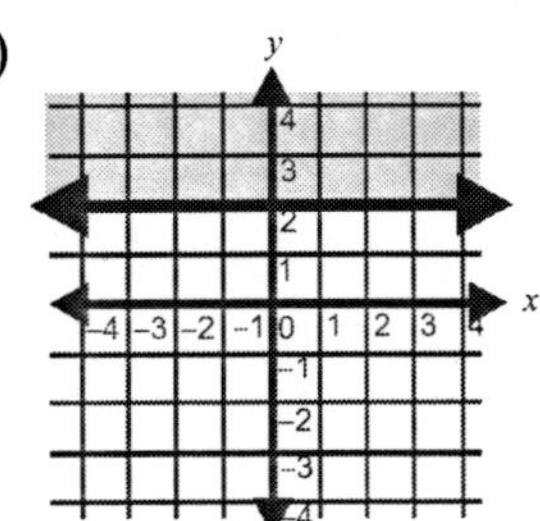

(B)

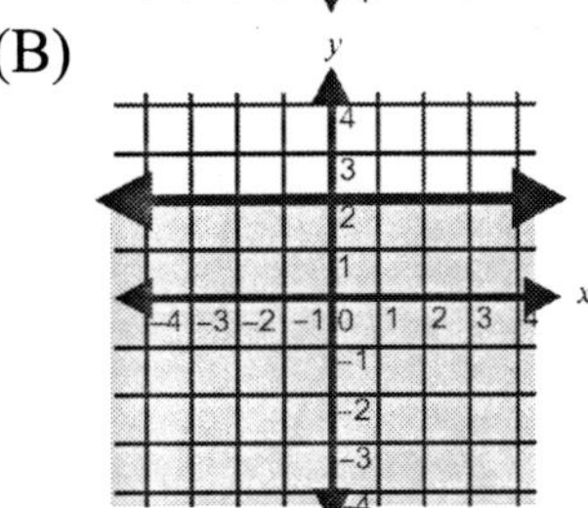

(C)

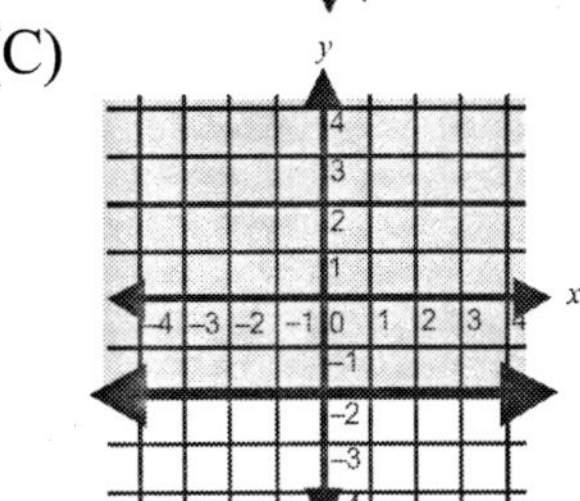

(D)

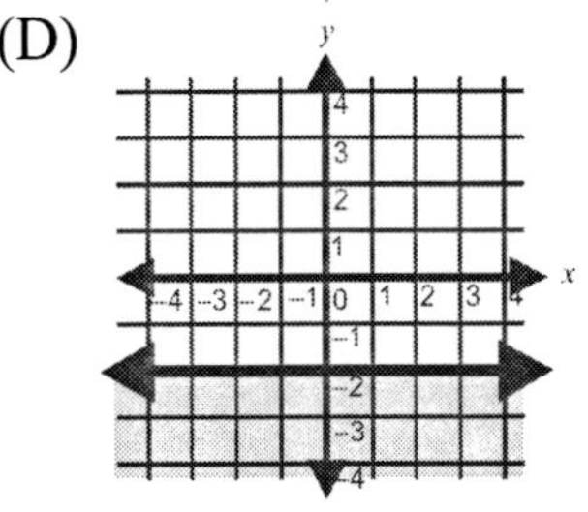

4.01a

36. Solve the equation $\sqrt{6w-8}=w$.

(A) $w=3,4$
(B) $w=2,4$
(C) $w=3\sqrt{2},4$
(D) $w=2\sqrt{3},-2\sqrt{3}$

4.02

37. Find: $(y^3-18y^2-5y)-(2y^2-5y+5)$.

(A) y^3-20y^2+5
(B) $y^3-20y^2-10y-5$
(C) $3y^3-16y^2-10y+5$
(D) y^3-20y^2-5

1.01b

38. Find the equation of a line perpendicular to the line shown below and passing through the point $(2, 1)$.

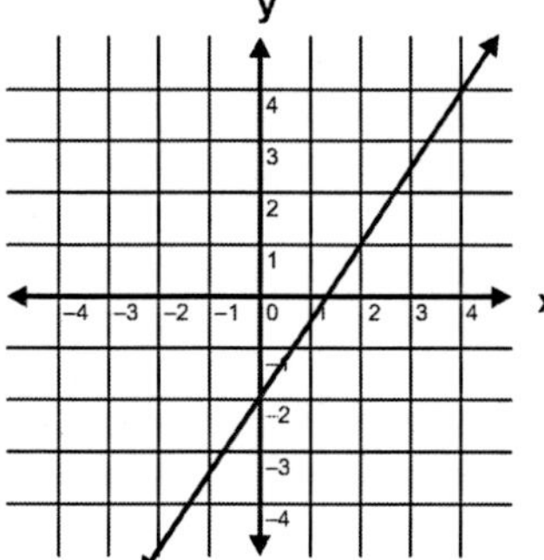

(A) $y = \frac{2}{3}x + \frac{7}{3}$

(B) $y = \frac{3}{2}x + \frac{4}{3}$

(C) $y = -\frac{2}{3}x + \frac{7}{3}$

(D) $y = \frac{3}{2}x + \frac{5}{2}$

2.02

39. Solve the equation $d^2 - 4d + 1 = 0$ by completing the square.

(A) $d = -3, -1$
(B) $d = \sqrt{3}, 2\sqrt{3}$
(C) $d = 2 - \sqrt{3}, \sqrt{3} + 2$
(D) $d = 2i, -2i$

4.02

40. Solve the equation $(x - 3)^2 = 1$.

(A) $x = 3, -3$
(B) $x = 1, 3$
(C) $x = 2, 4$
(D) $x = 1, -1$

4.02

41. What is the slope of the equation?

$3x - 3y = 5$

(A) -1
(B) 3
(C) -3
(D) 1

4.01b

42. Which is the graph of $2x - y = 1$?

(A)

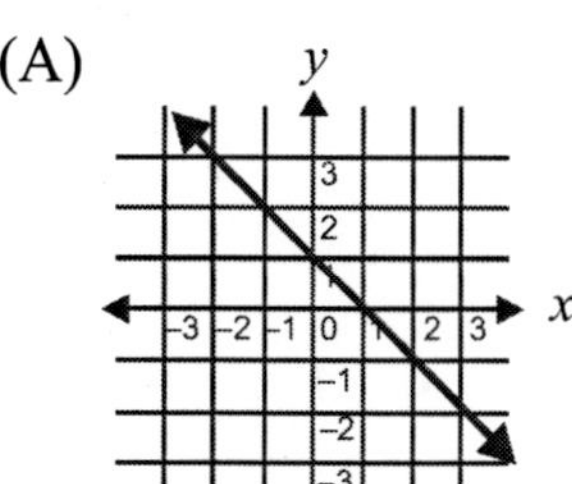

(B)

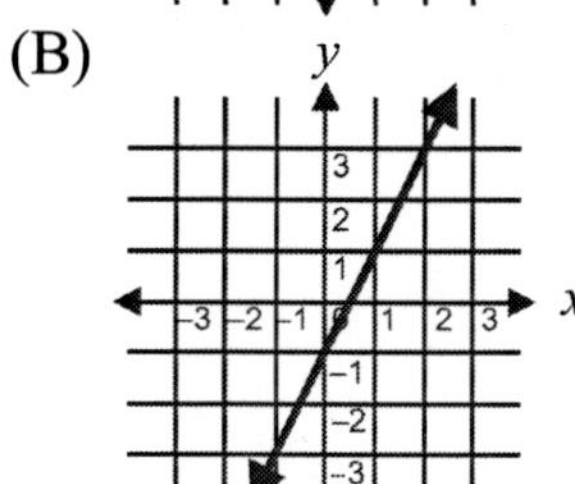

(C)

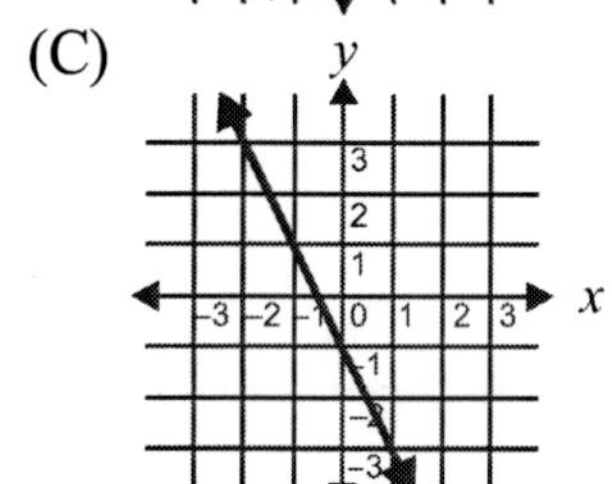

(D)

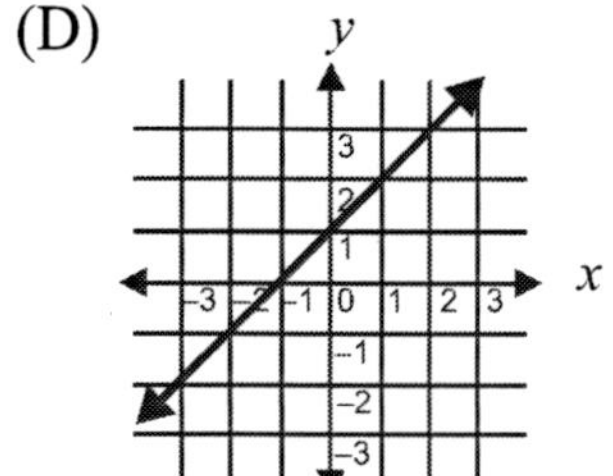

4.01a

43. Doug buys oak spindle chairs wholesale by purchasing 42 of them for $\$1,680.00$. To sell the chairs at a profit, Doug sells them in dining room sets of six with a solid oak table. Which formula will calculate Doug's cost for the chairs, x, in each dining room set?

(A) $42x = \$1,680$

(B) $\dfrac{\$1,680}{6} = x$

(C) $\dfrac{42}{6}x = \$1,680$

(D) $6x = \dfrac{\$1,680}{42}$

1.02

44. Identify the graph of the following function:

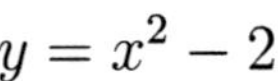

$$y = x^2 - 2$$

(A)

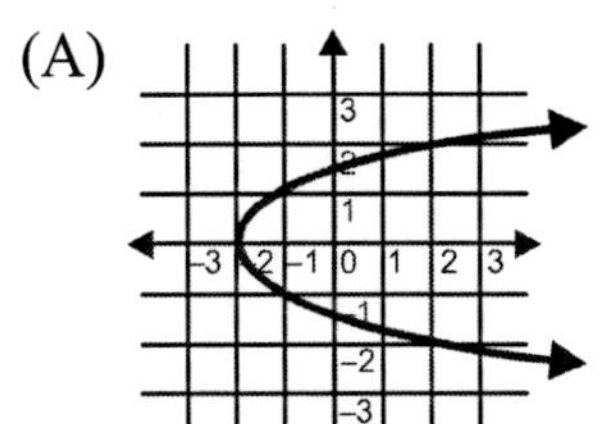

(B)

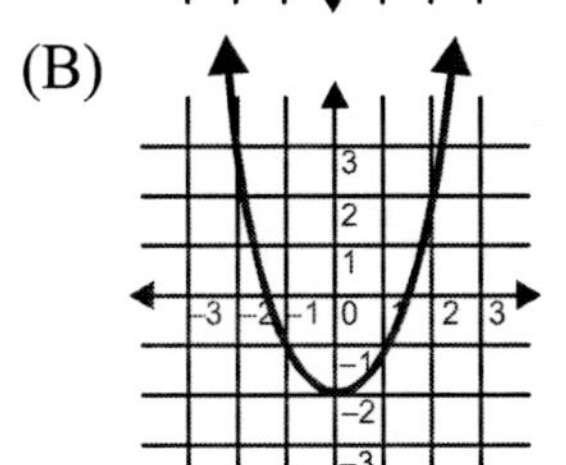

(C)

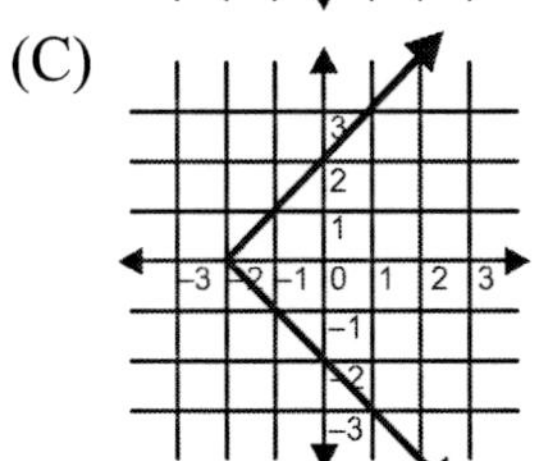

(D)

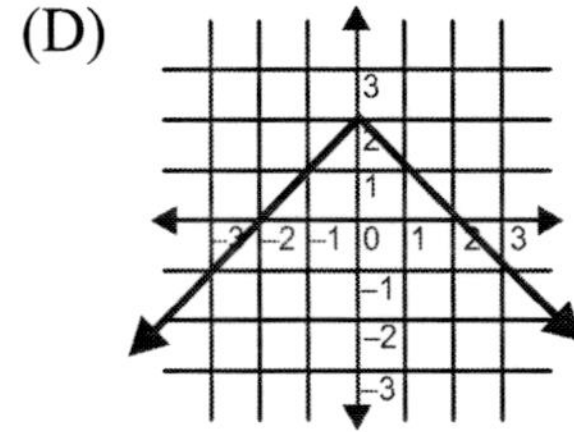

4.02

45. What is the equation of the line that includes the point $(3, -1)$ and has a slope of 2?

(A) $y = -2x - 7$
(B) $y = -2x - 2$
(C) $y = -2x + 7$
(D) $y = 2x - 7$

4.01b

46. Which is the graph of $-2x - 2y = -2$?

(A)

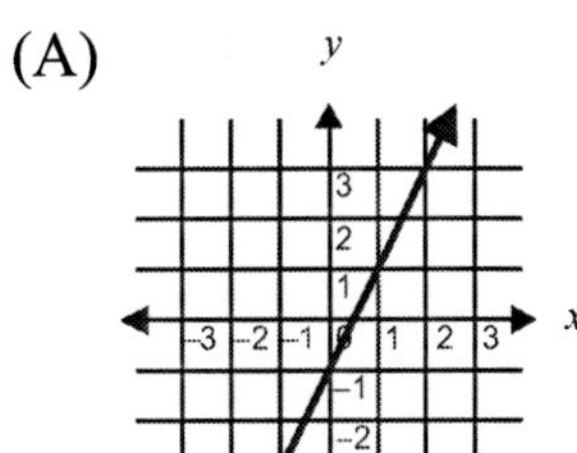

(B)

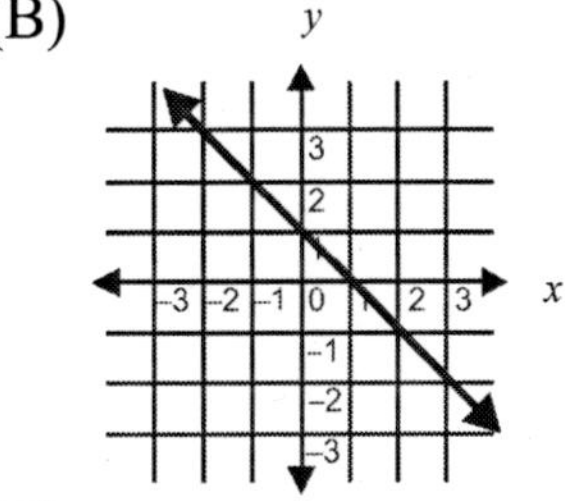

(C)

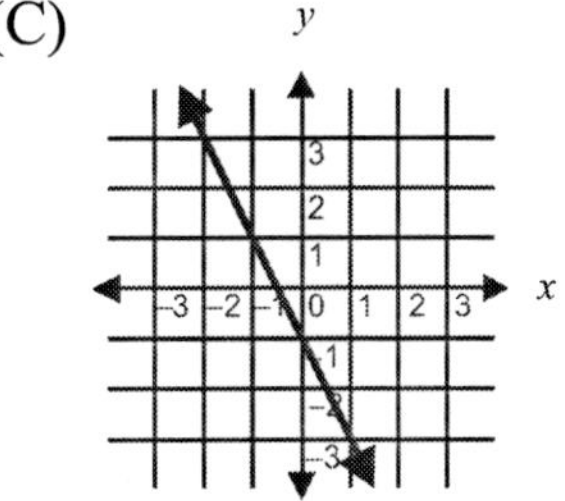

(D)

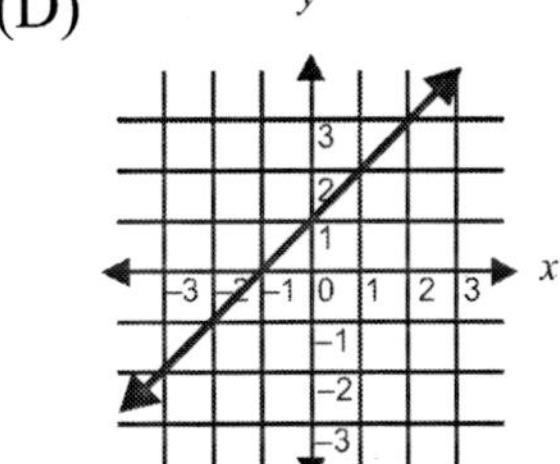

4.01a

47. Nicole works as an assistant pharmacist. She is paid \$9.40 per hour for the first 40 hours per week with time-and-a-half for overtime. Which equation would be used to determine her salary (s) where r is her regular hours, and v is her overtime hours work?

(A) $s = \$9.40(r + v) + .5v$
(B) $s = \$9.40r + 1.5(\$9.40)v$
(C) $s = 40r + 1.5(\$9.40)v$
(D) $s = 40r + 1.5v$

4.01b

48. Today Emily has driven 180 miles in 4 hours. Tomorrow she wants to drive 275 miles. Approximately how many hours should Emily plan to drive tomorrow?

(A) 2
(B) $5\frac{1}{2}$
(C) 6
(D) 7

1.03

49. The illustration below shows the function $f(x) = -x$.

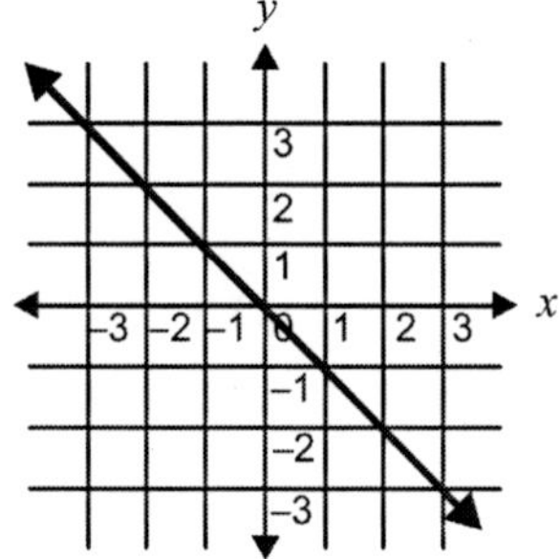

Which graph below shows the function $f(x) = -2x$?

(A)
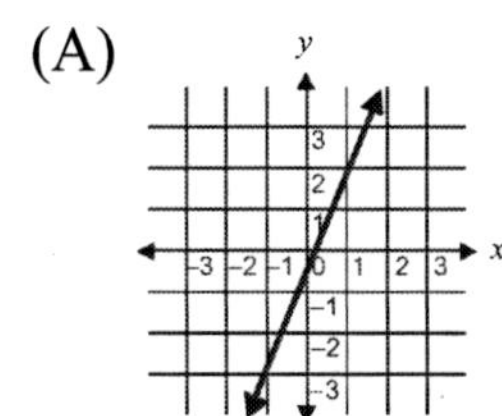
(B)
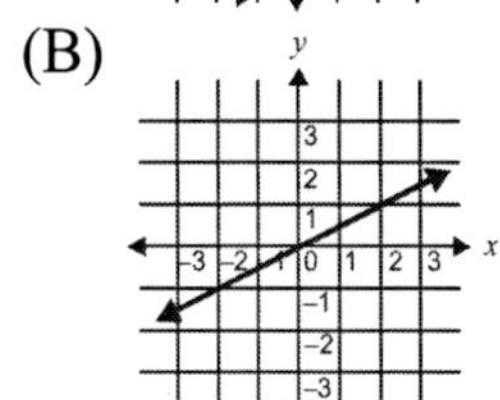
(C)
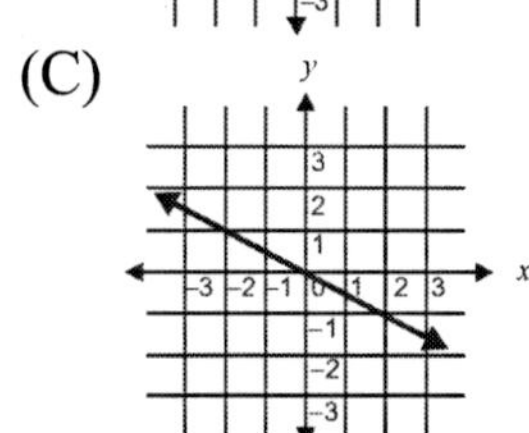
(D)
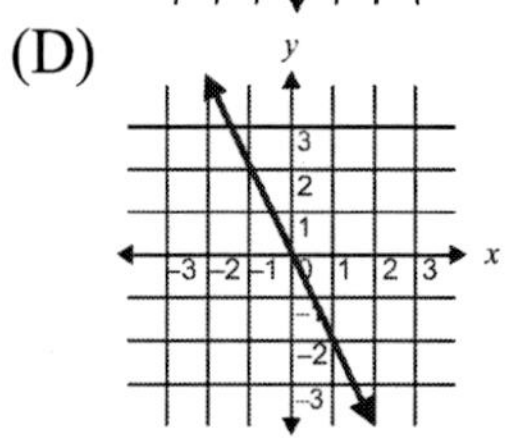

4.01b

50. Multiply: $6w(-5w^3 + 2w^2 - 4w)$

(A) $-30w^3 + 12w^2 - 24w$
(B) $-5w^3 + 2w^2 + 2w$
(C) $-30w^4 + 12w^3 - 24w^2$
(D) $-30w^4 + 12w^3 + 24w^2$

1.01b

51. Water hyacinths were introduced into the swamps of Louisiana to put oxygen back into the water. The hyacinths reproduced rapidly and soon became a nuisance. Read the growth table below, and then answer the question that follows.

Number of Days	Number of Hyacinths
1	2
21	4
41	8
61	16

Assuming the pattern continues, how many hyacinths will there be at 121 days?

(A) 32
(B) 64
(C) 128
(D) 256

4.04

52. Find the point of intersection of the two equations by adding and/or subtracting.

$x + y = 4$
$2x - y = 5$

(A) $(3, 1)$
(B) $(-3, 1)$
(C) $(1, 3)$
(D) $(-1, -3)$

4.03

53. What is the range of the function $x^2 + 2$ as represented in the graph?

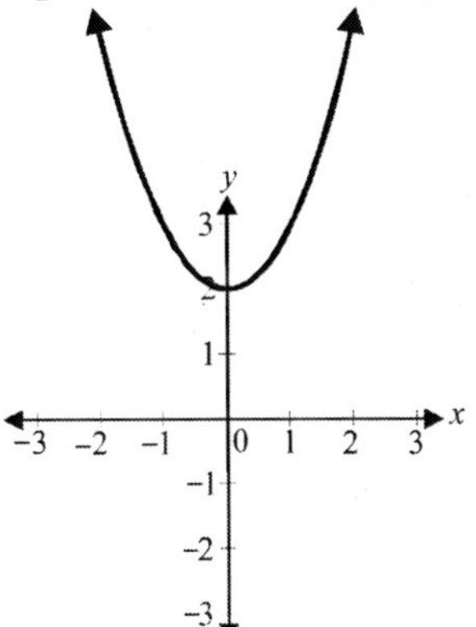

(A) $-3 \leq R \leq 3$
(B) $R \geq 0$
(C) All real numbers ≥ 2
(D) All real numbers

4.02

54. Find the point of intersection of the two equations.

$4x + 5y = \frac{2}{3}$
$7x - 3y = 9$

(A) $\left(\frac{2}{3}, -1\right)$
(B) $\left(-2, \frac{3}{2}\right)$
(C) $(-1, 0)$
(D) $\left(1, -\frac{2}{3}\right)$

4.03

55. What is the slope of the equation $3x - y = 5$?

(A) -1
(B) 3
(C) -3
(D) 1

4.01b

56. Solve the following quadratic equation by factoring. $6x^2 - 16x - 6 = 0$

(A) $2, -3$
(B) $3, -\frac{1}{3}$
(C) $-1, 1$
(D) $2, -\frac{1}{3}$

4.02

57. For this graph of a quadratic function in the form of $y = ax^2 + c$, what are the values of a and c?

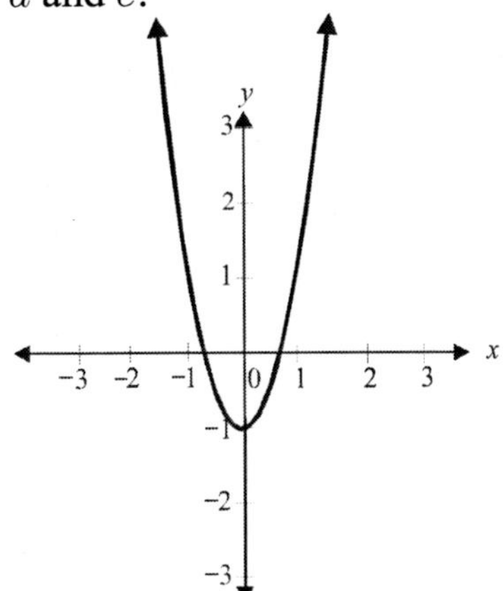

(A) $a = 1, c = 1$
(B) $a = -1, c = -1$
(C) $a = 2, c = -1$
(D) $a = 2, c = 1$

4.02

58. Mrs. Gilliam went to the Green's Nursery to buy bulbs to plant in her yard. She bought some on Friday and went back and bought more bulbs on Saturday. The table below shows what she bought and the amount she paid.

	Tulips	Daffodils	Total Cost
Friday	6	14	\$8.50
Saturday	20	7	\$14.45

What was the cost of one tulip?

(A) \$0.35
(B) \$0.40
(C) \$0.60
(D) \$0.65

4.03

59. Solve: $7 - 3(4x + 6) = 1$

(A) $x = 5$
(B) $x = -1$
(C) $x = 2$
(D) $x = 1$

4.01a

60. After the Valentine's Day party, Jenny, Luisa, and Carla had 68 pieces of candy combined. Jenny got 4 more pieces of candy than Carla, and Luisa got 6 more pieces of candy than Jenny. How many pieces did Carla get?

(A) 22
(B) 29
(C) 18
(D) 23

4.03

61. Which of the following is the graph of $\frac{1}{3}y = -\frac{2}{3}x + 1$?

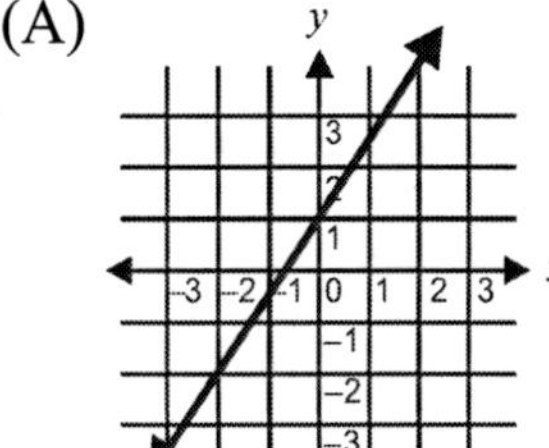

(B)

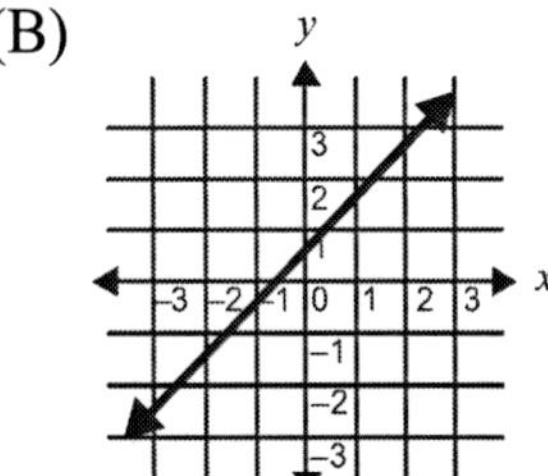

(C)

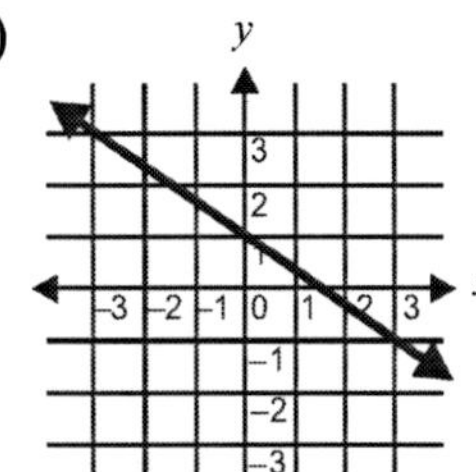

(D)

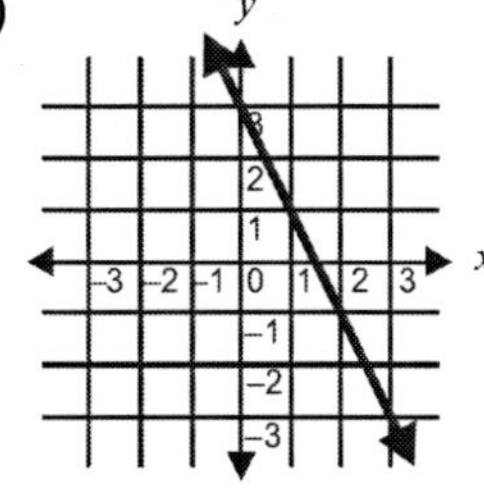

4.01a

62. Periwinkles are snails that live in tidepools near the shores of the ocean. A study by Jane Lubchenco shows the relationship between the number of periwinkles in a tidepool and the number of species of algae present in the same tidepool.

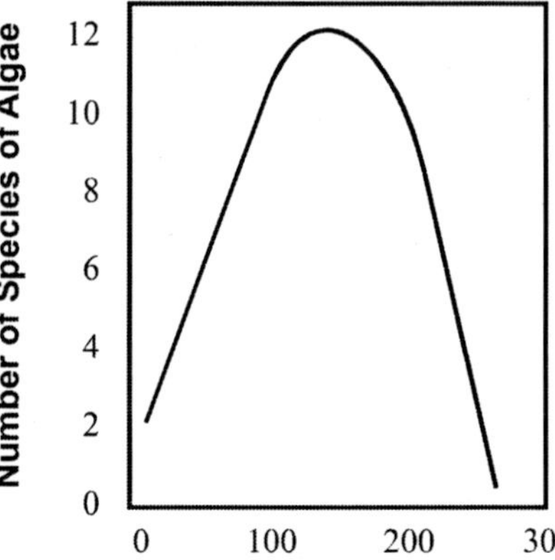

Which of the following statements best describes the relationship between the number of species of algae and the number of periwinkles present in the tidepool?

(A) The number of algae species is greatest when the number of periwinkles is in the intermediate range.
(B) As the number of periwinkles increases, the number of algae species increases.
(C) As the number of periwinkles increases, the number of algae species decreases.
(D) The number of algae species rises in direct proportion to increases in the number of periwinkles.

4.02

63. What are the roots of the quadratic equation below?

$x^2 - 2x - 24$

(A) $\{-8, 6\}$
(B) $\{-6, 4\}$
(C) $\{6, -4\}$
(D) $\{-8, 3\}$

4.02

64. Which of the following charts would be most useful in predicting height as a function of finger length?

(A)

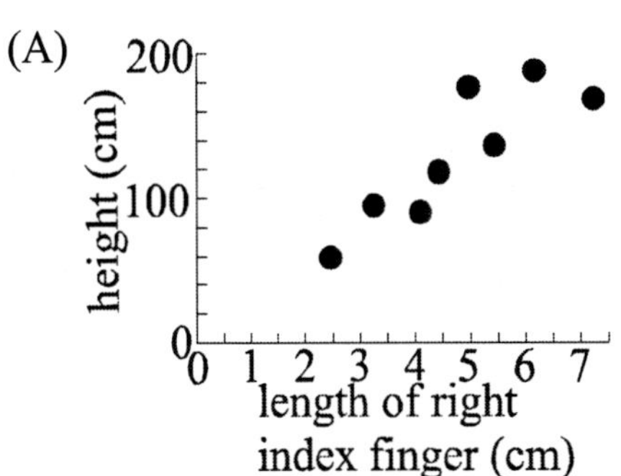

(B)

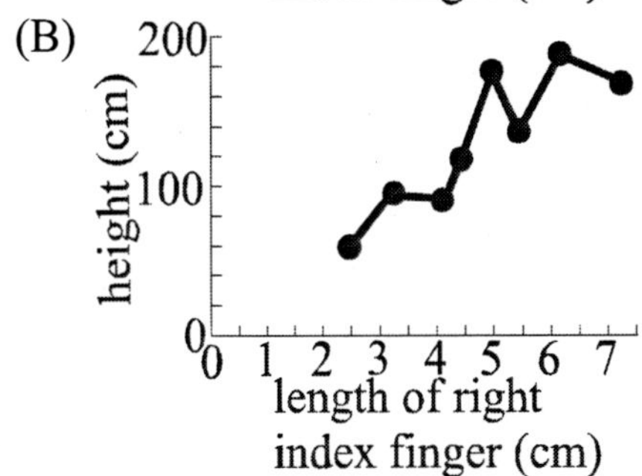

(C)

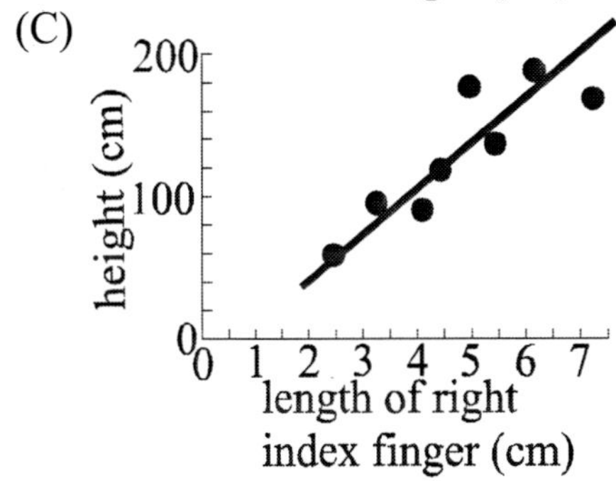

(D)

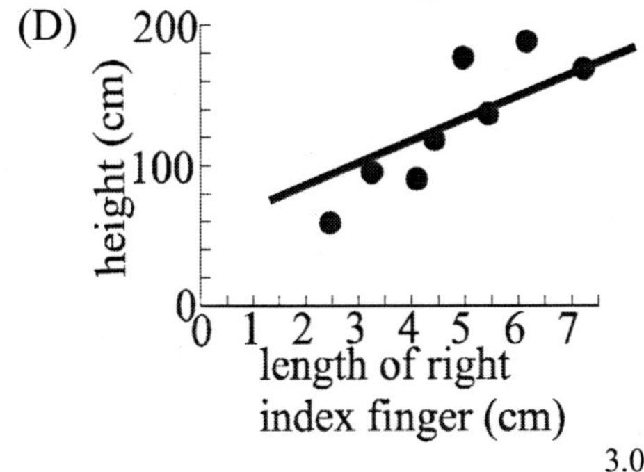

3.03b

65. The graph represents the equation $2y = 3x - 6$.

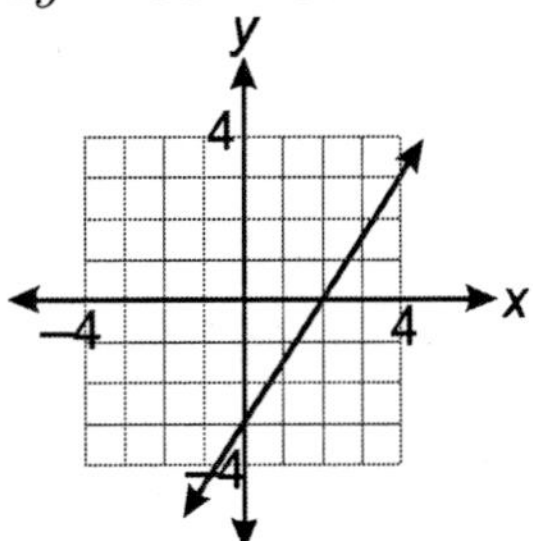

If the constant term changes from -6 to 4, what will the graph look like?

(A)

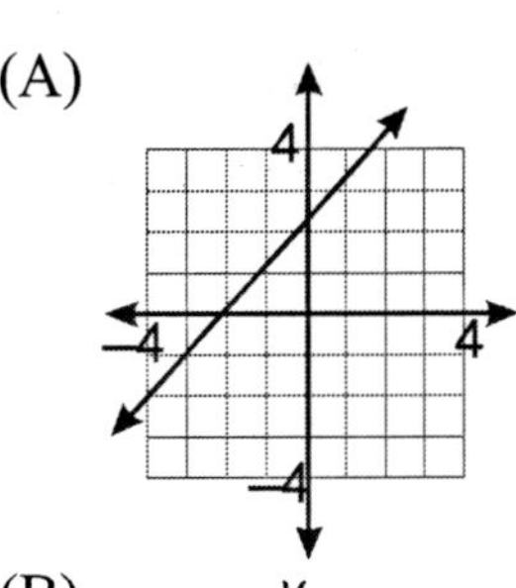

(B)

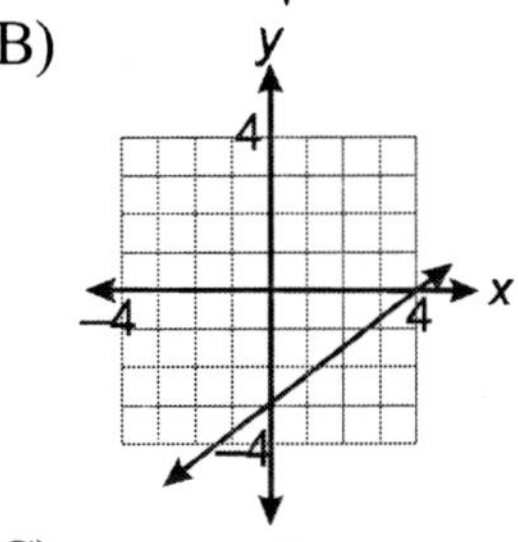

(C)

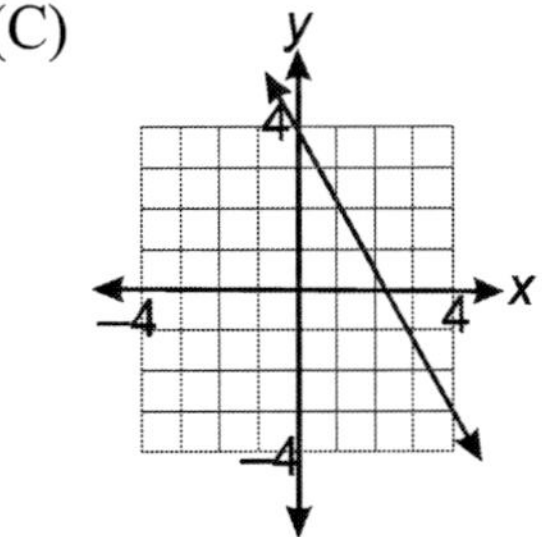

(D)

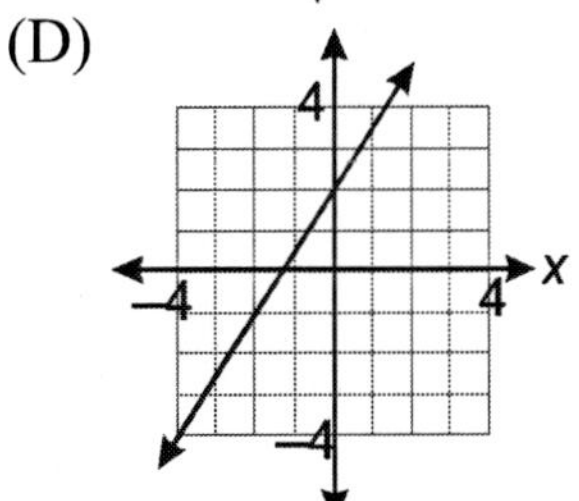

4.01b

66. Brice can type 22 pages in 3 hours. At this rate approximately how long would it take to type one page?

(A) 7.3 minutes
(B) 8.2 minutes
(C) 12.2 minutes
(D) 13.6 minutes

1.03

67. When a scientist pours more crystals in a jar every five minutes, the volume of the crystals triples. The jar becomes completely full of crystals at 3:47 PM. At what time was the jar only $\frac{1}{27}$ of the way full?

(A) 3:20 PM
(B) 3:22 PM
(C) 3:32 PM
(D) 3:37 PM

4.04

68. The functional relationship between altitude (A) above sea level (in feet) and the approximate boiling point (B) of water (in degrees Fahrenheit) may be expressed by the equation $B = -0.00176A + 212$. What is the approximate boiling point of water at $2,500$ feet?

(A) 194.4
(B) 207.6
(C) 208.3
(D) 216.4

4.01a

69. The Coffee Cottage has various sizes of coffee mugs. The cost (c) of a mug of coffee is based on the equation $c = 0.15q + 0.25$, where q is the number of ounces of coffee the mug holds. What is the slope (m) of the equation $c = 0.15q + 0.25$?

(A) $m = 0.10$
(B) $m = 0.15$
(C) $m = 0.25$
(D) $m = 0.40$

4.01b

70. What system of inequalities defines the shaded areas below?

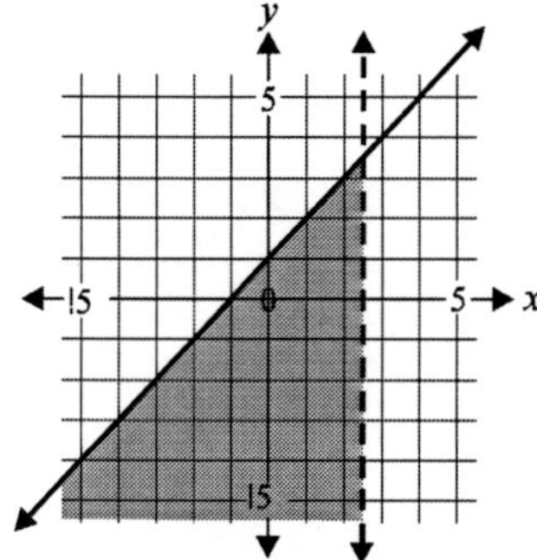

(A) $y \leq x + 1$
$x < 2.5$
(B) $y \geq x + 1$
$x > 2.5$
(C) $y \leq x + 1$
$x > 2.5$
(D) $y > x + 1$
$x < 2.5y$

4.03

71. Where is the y-intercept for the line $3x + 2y + 8 = 0$?

(A) $(2, 0)$
(B) $(0, 8)$
(C) $(0, -4)$
(D) $(-4, 0)$

4.01b

72. Factor $9x^2 + 15x - 14$

(A) $(9x - 1)(x + 7)$
(B) $(3x + 2)(3x - 7)$
(C) $(3x - 2)(3x - 7)$
(D) $(3x - 2)(3x + 7)$

1.01c

73. The coordinates of the endpoints of a line segments are $(3, 1)$ and $(-5, 9)$. Find the coordinates of the midpoint of the line segment.

(A) $(4, 5)$
(B) $(-1, 5)$
(C) $(-2, 4)$
(D) $(-2, 8)$

2.01

74. For the following pair of equations, find the point of intersection using the substitution method.

$-3x - y = -2$
$5x + 2y = 20$

(A) $(2, -4)$
(B) $(2, 5)$
(C) $(-16, 50)$
(D) $\left(\frac{1}{5}, \frac{1}{2}\right)$

4.03

75. $4\begin{bmatrix} 5 & 3 \\ -4 & 2 \\ 6 & 0 \end{bmatrix} - \begin{bmatrix} 10 & 7 \\ -4 & -8 \\ 1 & 4 \end{bmatrix} =$

(A) $\begin{bmatrix} 10 & 5 \\ -12 & 16 \\ 23 & -4 \end{bmatrix}$

(B) $\begin{bmatrix} 20 & 12 \\ -16 & 8 \\ 24 & 0 \end{bmatrix}$

(C) $\begin{bmatrix} -20 & -16 \\ 0 & 40 \\ 20 & -16 \end{bmatrix}$

(D) $\begin{bmatrix} 10 & -4 \\ 0 & 10 \\ 5 & -4 \end{bmatrix}$

3.02

76. $\frac{1}{2}\begin{bmatrix} 5 & 8 & -4 \\ -1 & 12 & 6 \end{bmatrix} =$

(A) $\begin{bmatrix} 2.5 & 8 & -4 \\ -1 & 12 & 6 \end{bmatrix}$

(B) $\begin{bmatrix} 10 & 16 & -8 \\ -2 & 24 & 12 \end{bmatrix}$

(C) $\begin{bmatrix} 2.5 & 4 & -2 \\ -\frac{1}{2} & 6 & 3 \end{bmatrix}$

(D) Not possible

3.02

77. Solve by factoring: $11x^2 - 31x - 6 = 0$.

(A) 3 and -2

(B) 3 and $-\frac{2}{11}$

(C) 3 and $-\frac{11}{2}$

(D) -3 and $\frac{11}{2}$

4.02

78. Using the following table and assuming that the population at the art show continues to increase during the first hour, how many people will be at the art show one hour after it starts?

TIME IN MINUTES	NUMBER OF PEOPLE AT THE ART SHOW
0	12
10	22
20	32
30	42

(A) 52
(B) 62
(C) 72
(D) None of the above

3.03a

79. Amin has a part-time job earning \$10.00 per hour. He made a chart of his hours, earnings, and federal taxes taken out of his paycheck.

Hours	Earned	Taxes
25	\$250	\$23
26	\$260	\$25
27	\$270	\$26
28	\$280	\$28
29	\$290	\$29

If the pattern continues, how much will be taken from his check for federal taxes if he works 32 hours?

(A) \$32
(B) \$33
(C) \$34
(D) \$35

3.03a

80. Carol is arranging rows of tables for discussion groups.

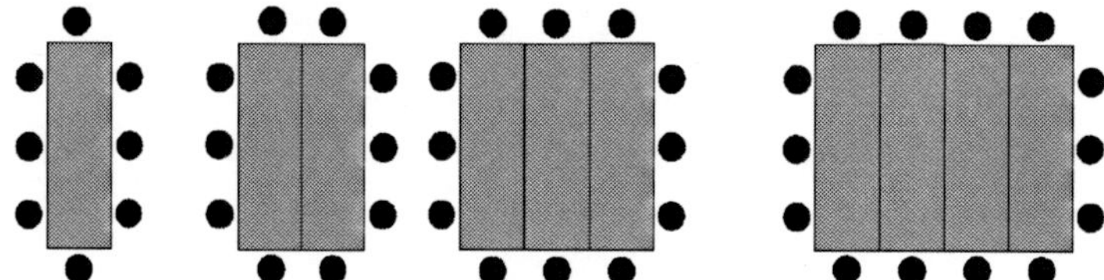

The table below shows the relationship between the number of tables (T) in a group and the maximum participants (P) in the group.

Number of Tables (T)	Maximum Participants (P)
1	8
2	10
3	12
4	14

Which of these functions generalizes the pattern of data in the table?

(A) $P = 2T + 6$
(B) $P = 3T + 2$
(C) $P = 3T + 5$
(D) $P = 4T + 2$ 3.03a

Practice Test 2

1. Della is renting a car for the day. The rental fee (y) is \$30 plus \$0.25 per mile (m). Which of the following equations represents this cost?

(A) $y = 0.30m + 25$
(B) $y = 30m + 0.25$
(C) $y = 0.25m + 30$
(D) $y = m(0.25 + 30)$

4.01b

2. $4^{-2} \times 2^{-3}$

(A) -48

(B) $\dfrac{1}{128}$

(C) $\dfrac{1}{48}$

(D) $\dfrac{3}{36}$

1.01a

3. Four students attempt to simplify a mathematical expression. They have four different answers. Which of the answers below is equivalent to the expression,

$2(a + 3b) - 4(3a - b) - (5a + 4b)$?

(A) $-15a + 6b$
(B) $-17a + 9b$
(C) $-9a - 2b$
(D) $-9a + 9b$

1.01b

4. $14(x - 6) = -26$

(A) $x = 58$
(B) $x = 4\frac{1}{7}$
(C) $x = -7\frac{6}{7}$
(D) $x = 29$

4.01a

5. Ryan is dieting. He has plotted his weight on the first of each month for the past 5 months on the graph below.

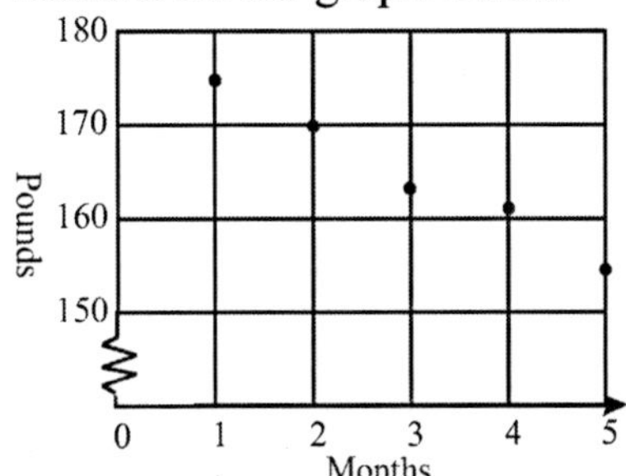

The slope of a straight line most nearly representing Ryan's weight as a function of month number would be approximately

(A) -10

(B) -5

(C) $-\frac{1}{2}$

(D) $\frac{1}{2}$

3.03a

6. Which of the following is an equation of a line that is perpendicular to the line l in the graph with a y-intercept of $(0, 2)$?

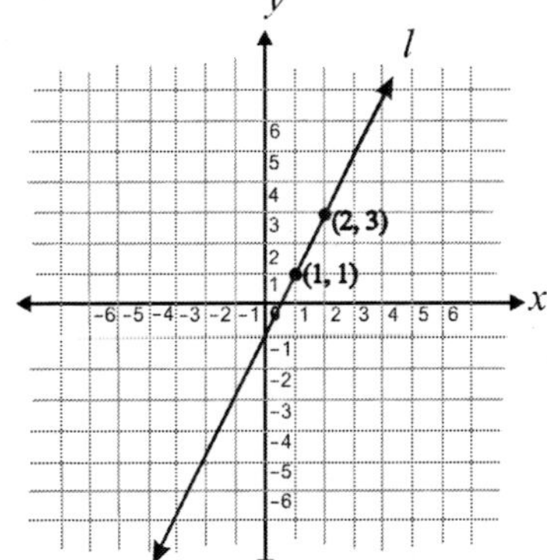

(A) $x - 2y = -4$
(B) $x - 2y = 4$
(C) $x + 2y = 4$
(D) $2x + y = 4$

2.02

7. Use the correct order of operations to evaluate the following expression.

$-3(x-5)^2$

(A) $-3x^2-10x+25$
(B) $x^2-10x+25$
(C) $3x^2-30x+75$
(D) $-3x^2+30x-75$

4.02

8. Simplify the following monomial.

$2 \cdot x^4 \cdot y^6 \cdot x^{-4}$

(A) $2y^6$
(B) $2(xy)^6$
(C) $64y^6$
(D) $2x^{-8}y^6$

1.01b

9. Peter buys T-shirts wholesale by purchasing 125 of them for \$250. To sell the shirts at a profit, Peter sells them with his own designs in sets of five. Which formula will calculate Peter's cost, x, for the shirts in each set?

(A) $\frac{5}{125}x = \$250$

(B) $\frac{\$250}{5} = x$

(C) $x = 5\left(\frac{\$250}{125}\right)$

(D) $125x = \$250$

1.02

10. Solve for x in the following equation.

$\frac{6x-19}{-2} = 3.5$

(A) 12
(B) $\frac{13}{3}$
(C) 2
(D) 4

4.01a

11. The illustration below shows the function $f(x) = 3x - 1$.

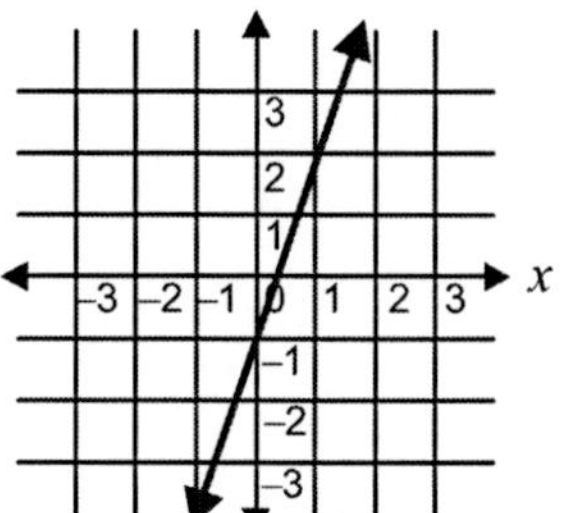

Which graph below shows the function $f(x) = -3x - 1$?

(A)

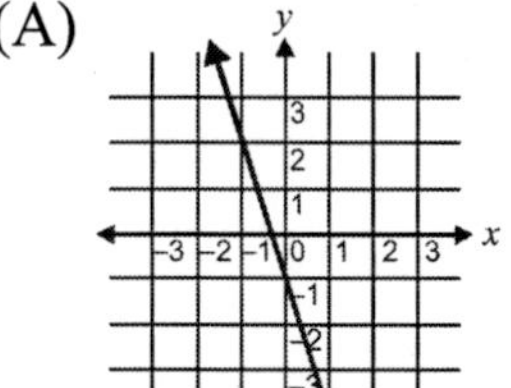

(B)

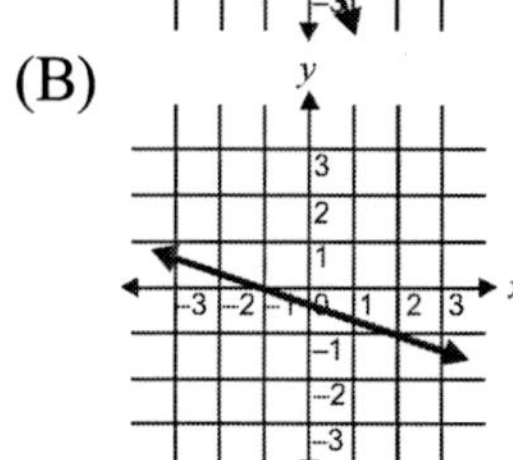

(C)

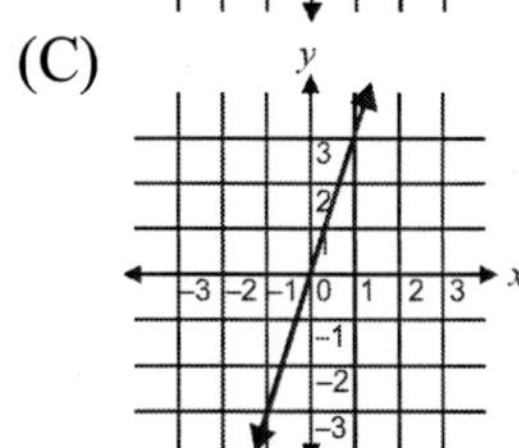

(D)

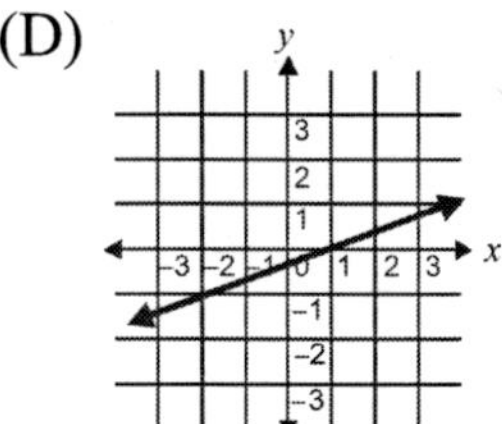

4.01b

12. Solve: $3(5x+3)+5(4x-9)=34$

(A) $x = 1$
(B) $x = 2$
(C) $x = -1$
(D) $x = -2$

4.01a

13. Solve $-4(2x+7) > 3(4x+5)+27$

(A) $x > \frac{7}{2}$

(B) $x < \frac{7}{2}$

(C) $x < -\frac{7}{2}$

(D) $x > \frac{1}{4}$

4.01a

14. If the equation below were graphed, which of the following points would lie on the line?

$x - 7y = 21$

(A) $(7, 3)$
(B) $(0, -3)$
(C) $(14, 0)$
(D) $(-3, 14)$

4.01a

15. Celeste earns $7.00 per hour for the first 40 hours she works this week and time-and-a-half for 5 hours of overtime. Her deductions total $74.82. Which equation will help Celeste figure her pay?

(A) $40(7) + 5(7 \times .5) - 74.82$
(B) $40(7+5) - 74.52$
(C) $45(7) - 74.82$
(D) $7\,[40 + 5(1.5)] - 74.82$

1.02

16. Simplify: $\frac{(2^3)^2}{(3)^{-1}}$

(A) $21\frac{1}{3}$
(B) 96
(C) 48
(D) 192

1.01a

17. Solve for x in the following equation.

$5x + 40 \leq 42$

(A) $x \geq \frac{2}{5}$

(B) $x \geq -\frac{2}{5}$

(C) $x \leq \frac{2}{5}$

(D) $x \leq -\frac{2}{5}$

4.01a

18. Find the x- and y- intercepts for the following equation: $2x + 5y = 30$.

(A) x-intercept $= 15$
y-intercept $= 6$
(B) x-intercept $= 5$
y-intercept $= 4$
(C) x-intercept $= 6$
y-intercept $= 15$
(D) x-intercept $= 4$
y-intercept $= 5$

4.01a

19. What are the x and y intercepts of the equation graphed below?

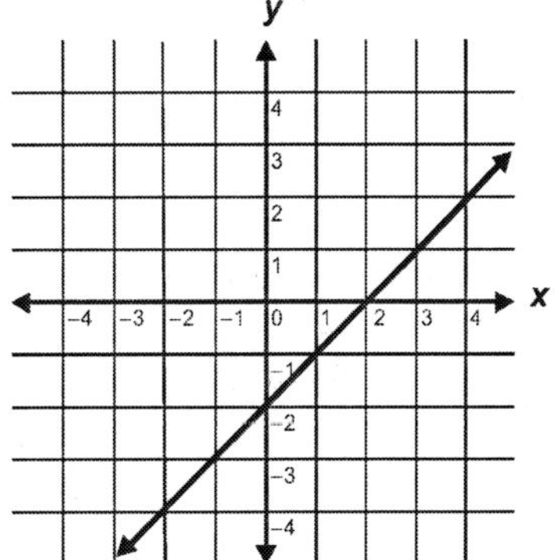

(A) x-intercept $= 1$; y-intercept $= -1$
(B) x-intercept $= -2$; y-intercept $= 2$
(C) x-intercept $= -1$; y-intercept $= 1$
(D) x-intercept $= 2$; y-intercept $= -2$

4.01a

20. The graph of which pair of equations below are parallel?

(A) $x + 4y = 3$
$3x + 4y = 3$
(B) $x - 4y = 3$
$4y - x = -3$
(C) $2x - 8 = 2y$
$2x + 8 = 2y$
(D) $6x + 6 = 6y$
$11x - 12 = 7y$

2.02

21. Which ordered pair is a solution for the following system of equations?

$-3x + 7y = 25$
$3x + 3y = -15$

(A) $(-13, -2)$
(B) $(-6, 1)$
(C) $(-3, -2)$
(D) $(-20, -5)$

4.03

22. Simplify: $3(5x - 2) + (-4x + 5)$

(A) $4x$
(B) $4x - 7$
(C) $11x - 11$
(D) $11x - 1$

1.02

23. Which of these is the equation that generalizes the pattern of the data in the table?

x	f(x)
–3	–5
–1	1
2	10
5	19

(A) $f(x) = 3x$
(B) $f(x) = x + 3$
(C) $f(x) = 2x + 6$
(D) $f(x) = 3x + 4$

3.03a

24. Solve: $\dfrac{x+4}{2} + 3 = 15$

(A) 2
(B) 20
(C) 32
(D) 40

4.01a

25. Which of these graphs represents the inequality $y \geq 2x + 1$?

(A)
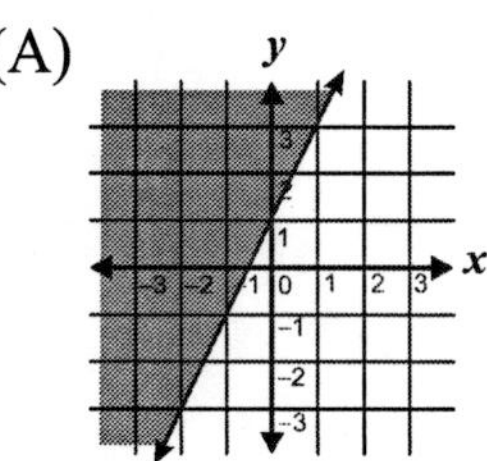
(B)
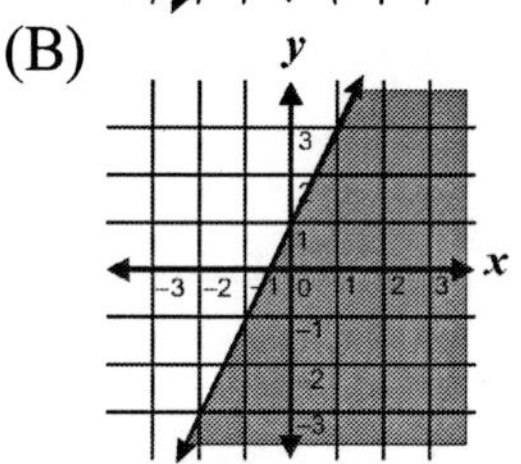
(C)
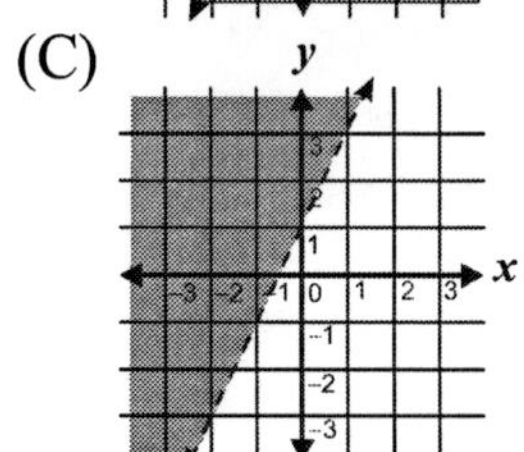
(D)
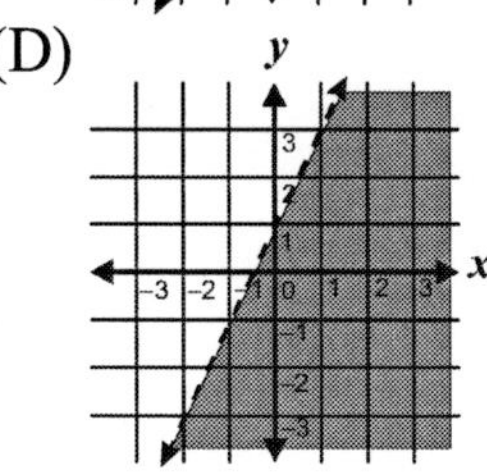

4.01a

26. Solve: $4(2x - 1) = x - 6(x + 3)$

(A) $-\frac{14}{13}$

(B) $\frac{7}{13}$

(C) $\frac{2}{3}$

(D) $\frac{4}{3}$

4.01a

27. Which of these graphs represents 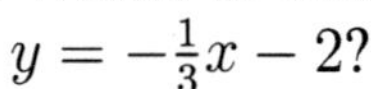 $y = -\frac{1}{3}x - 2$?

(A)

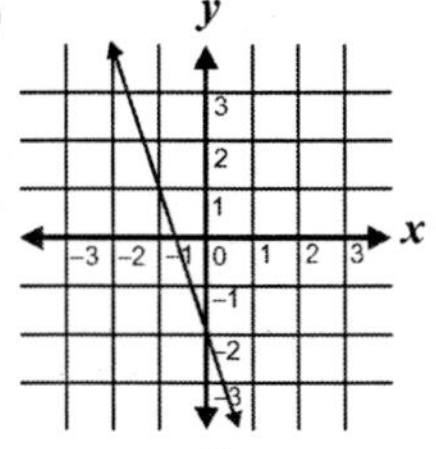

(B)

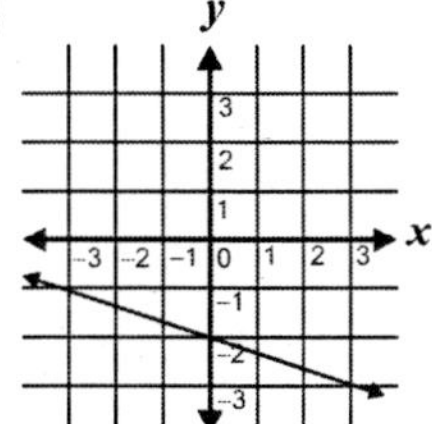

(C)

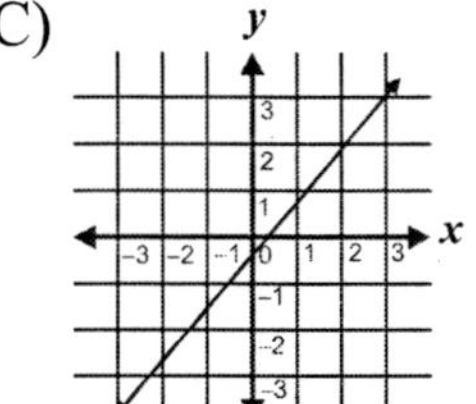

(D)

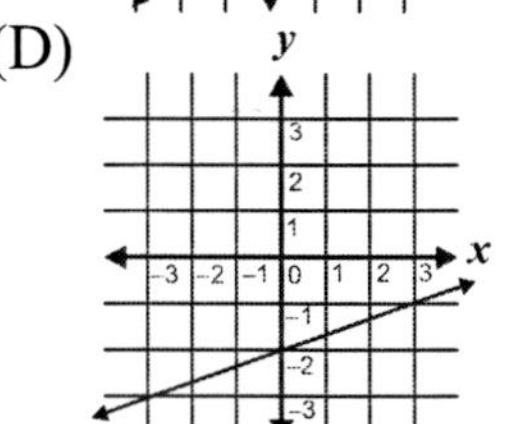

4.01a

28. For the following pair of equations, find the point of intersection.

$$3x + 3y = 9$$
$$9y - 3x = 6$$

(A) $\{1, 2\}$

(B) $\left\{\frac{7}{4}, \frac{5}{4}\right\}$

(C) $\{1, 1\}$

(D) $\left\{\frac{1}{3}, \frac{1}{6}\right\}$

4.03

29. The graph represents the equation $2y = x - 4$.

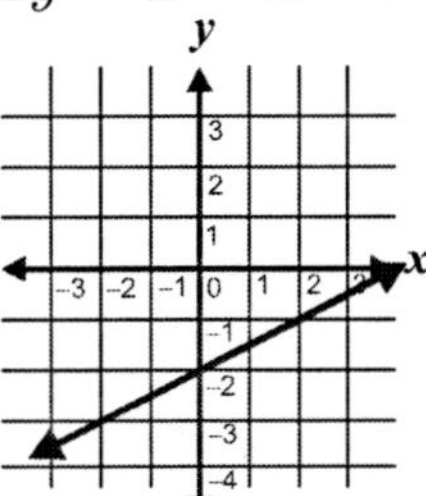

If the coefficient of y changes from 2 to 1, what will the graph look like?

(A)

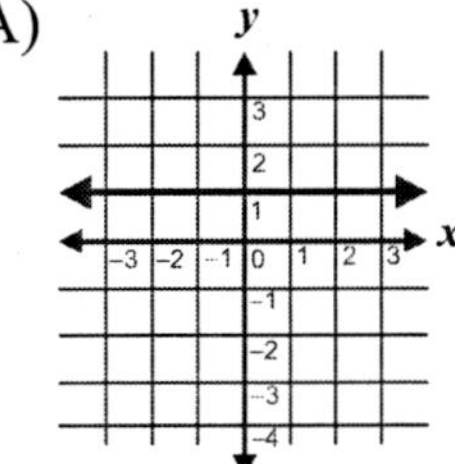

(B)

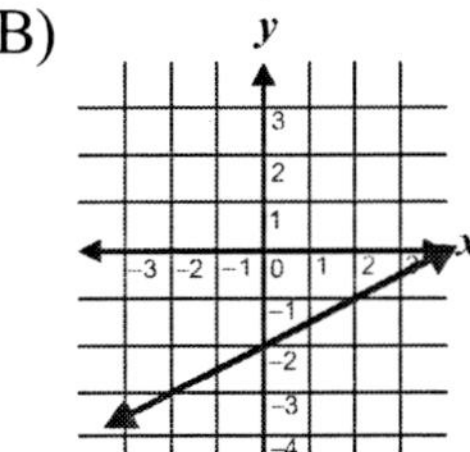

(C)

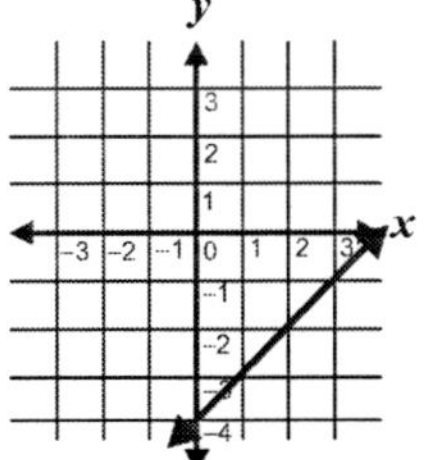

(D)

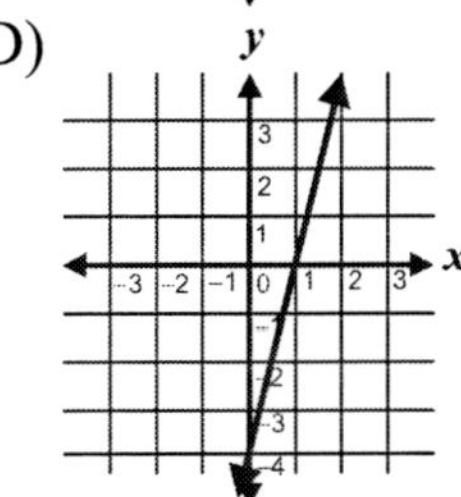

4.01b

30. Simplify: $(5x - 4)(x + 2)$

(A) $5x^2 - 2$
(B) $5x^2 - 8$
(C) $5x^2 + 6x - 8$
(D) $5x^2 - 18x - 8$

4.02

31. Solve $ab + cd = 20$ for b.

(A) $b = 20 - cda$

(B) $b = \dfrac{20 + cd}{a}$

(C) $b = \dfrac{20 - cd}{a}$

(D) $b = 20cda$

4.01a

32. Find the equation of the line perpendicular to the line graphed below with the same y-intercept.

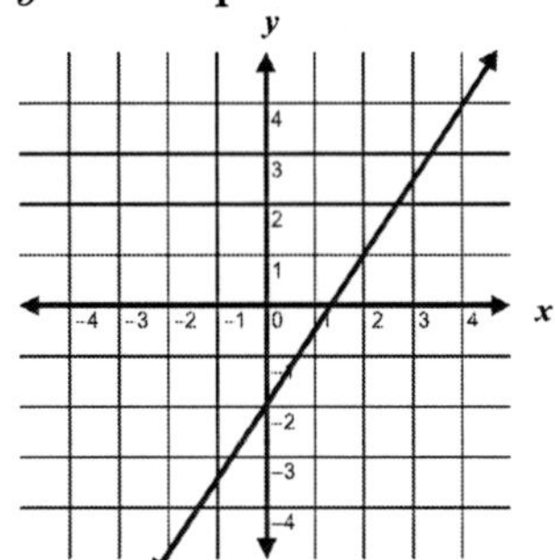

(A) $y = -2x - 2$

(B) $y = -\frac{2}{3}x - 2$

(C) $y = -\frac{3}{2}x + 2$

(D) $y = \frac{3}{2}x - 2$

4.01a

33. Which equation is non-linear?

(A) $y = \frac{1}{4}x + 2$
(B) $y = -x^2$
(C) $x + 2y = -4$
(D) $2x - 4 = 0$

4.02

34. Solve the equation $(x + 9)^2 = 49$

(A) $x = -9, 9$
(B) $x = -9, 7$
(C) $x = -16, -2$
(D) $x = -7, 7$

4.02

35. Which graph shows a line with a slope of $\frac{5}{2}$, passing through the point $(1, -3)$?

(A)

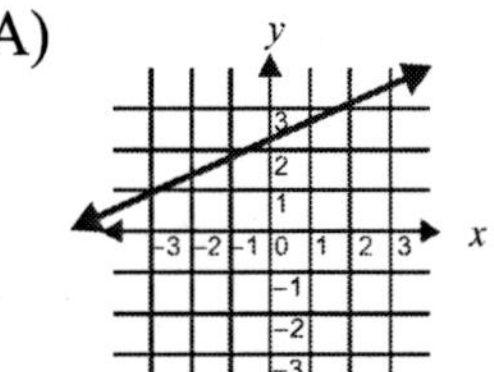

(B)

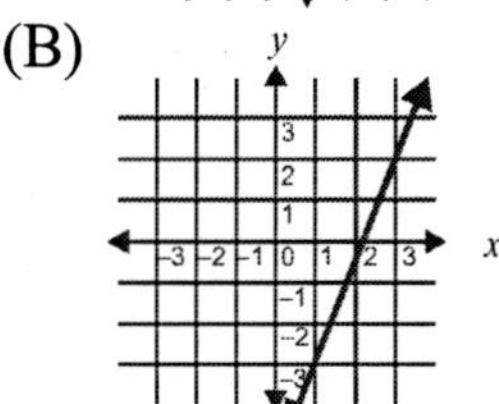

(C)

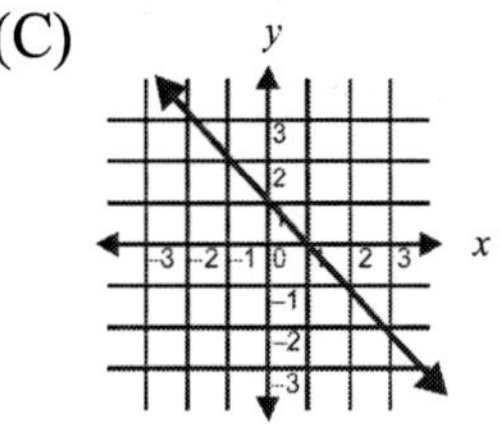

(D)

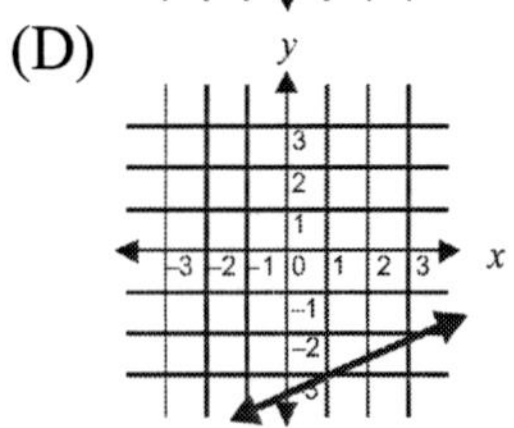

4.01b

36. Solve: $6a^2 + 11a - 10 = 0$

(A) $\left\{-\frac{2}{5}, \frac{3}{2}\right\}$

(B) $\left\{\frac{2}{5}, \frac{2}{3}\right\}$

(C) $\left\{-\frac{5}{2}, \frac{2}{3}\right\}$

(D) $\left\{\frac{5}{2}, \frac{2}{3}\right\}$

4.02

37. Solve: $1 + 3x - 9 = 4x - 7$

(A) -15

(B) $-\frac{15}{7}$

(C) -1

(D) $-\frac{7}{15}$

4.01a

38. Solve the equation $y = \sqrt{16 - 6y}$

(A) $y = 4, 2\sqrt{3}$
(B) $y = 3 - \sqrt{2}, 2 - \sqrt{3}$
(C) $y = 2, -8$
(D) $y = 2 + 3i, 2 - 3i$ 4.02

39. Solve the equation $c^2 + 3c - 9 = 0$ by completing the square.

(A) $c = 3, -3$
(B) $c = \frac{3}{2}\sqrt{5} - \frac{3}{2}, -\frac{3}{2}\sqrt{5} - \frac{3}{2}$
(C) $c = \pm\sqrt{3}$
(D) $c = 3i, -3i$ 4.02

40. What is the slope of the equation $-2x^2 + 4y = 7$?

(A) $\frac{1}{2}$
(B) 2
(C) -2
(D) $-\frac{1}{2}$ 4.02

41. Find: $(3y^3 + 5y^2 - 8) + (4y^3 - 6y^2 + 3)$

(A) $-y^3 + 5y^2 - 6y - 5$
(B) $7y^3 - y^2 - 5$
(C) $7y^3 + 11y^2 - 5$
(D) $-y^3 + 7y^2 - 6$ 1.01b

42. Multiply: $-6a^3(-2ab^2 + 5a^2b - 6a^3)$

(A) $12a^3b^2 - 30a^6 + 36a^9$
(B) $12a^4b^2 - 30a^5b + 36a^6$
(C) $-12a^3b^2 + 30a^6 - 36a^9$
(D) $-12a^4b^2 + 30a^5b - 36a^6$ 4.01b

43. Solve: $-\frac{4}{5}x \geq 8$

(A) $x \geq 10$
(B) $x \geq 5$
(C) $x \leq -10$
(D) $x \geq 10$ 4.01a

44. Which of the following is a graph of the inequality $x + y \geq 4$?

(A)
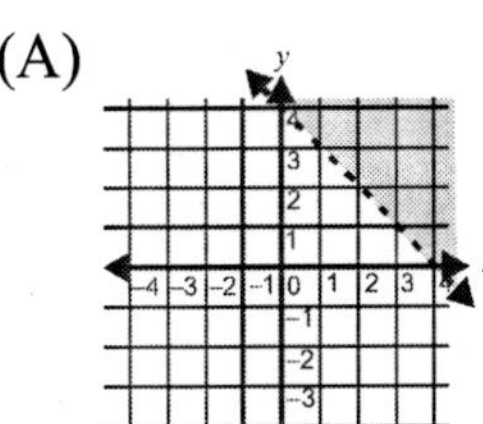

(B)
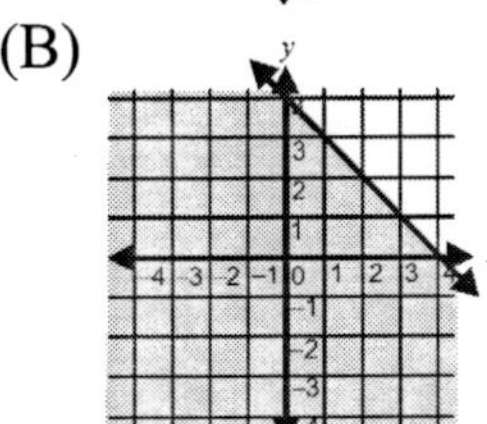

(C)
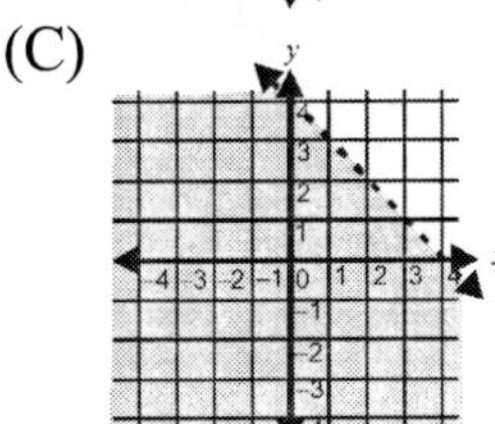

(D)
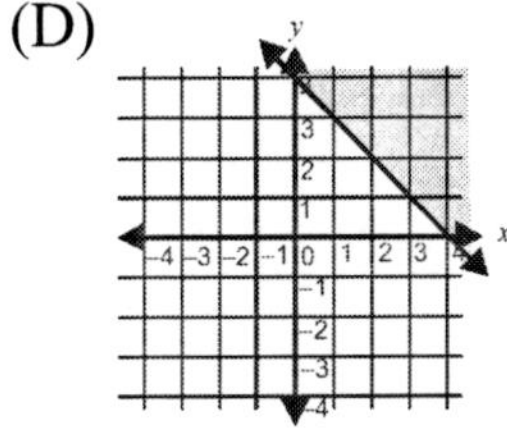

4.01a

45. Solve for x in the following equation: $x + 2(x + 200) + 800 = 3000$

(A) $1,200$
(B) 800
(C) 600
(D) $1,600$

4.01a

46. Factor: $b^2 - 2b - 8$

(A) $(b - 4)(b + 4)$
(B) $(b - 2)(b + 4)$
(C) $(b + 2)(b - 4)$
(D) $(b - 2)(b - 2)$

1.01c

47. Which ordered pair is a solution for the following system of equations?

$5x + 7y = -15$
$4x - 6y = 46$

(A) $(4, -5)$
(B) $(-3, 0)$
(C) $(5, -5)$
(D) $(-1, -7)$

4.03

48. Which of the following relations is a function?

(A) $\{(-2, 2)(-2, 1)(0, 0)(-1, -1)\}$
(B) $\{(4, 9), (2, 8)(3, 7)(6, 5)\}$
(C) $\{(1, 5)(1, 4)(1, 3)(1, 2)\}$
(D) $\{(0, 1)(2, 4)(2, 6)(3, 9)\}$

3.03b

49. Veronica is comparing the cost of two floor waxing services. Acme Co. charges \$50 per 100 square feet. Best, Inc. charges \$25 per 100 square feet plus a \$75 fee. Two equations representing price as a function of square footage are graphed below.

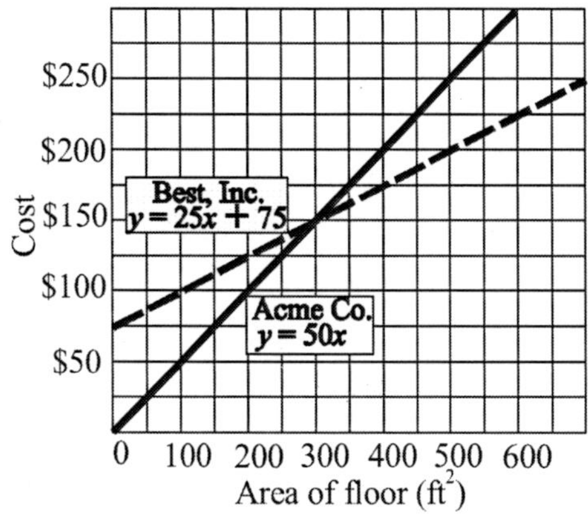

Best, Inc. is less expensive than Acme for

(A) all size floors.
(B) exactly 300 ft^2.
(C) areas less than 300 ft^2.
(D) areas more than 300 ft^2.

4.03

50. Sherry wants to find out if height is correlated to weight. She asks 10 people what their height and weight is and the data is listed below.

(63″, 150 lbs), (67″, 160 lbs), (72″, 190 lbs), (62″, 180 lbs), (64″, 130 lbs), (70″, 210 lbs), (63″, 120 lbs), (68″, 165 lbs), (65″, 145 lbs), and (68″, 220 lbs)

Which graph represents this data with an accurate line of best fit?

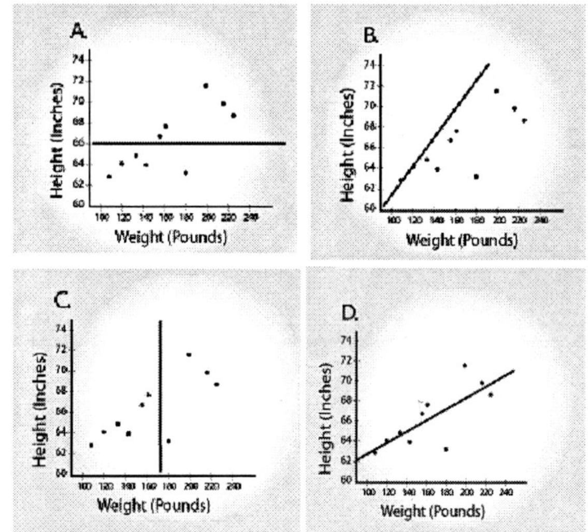

(A) Graph A
(B) Graph B
(C) Graph C
(D) Graph D

3.03b

51. In a survey of 125 randomly selected voters, 80 said they would vote for Steven Gillmor. If 20,000 people in the district vote in the election, approximately how many would be expected to vote for Steven Gillmor?

(A) 3,125
(B) 12,800
(C) 14,875
(D) 16,000

1.03

52. Mileage Between Cities

	Asheville	Charlotte	Durham	Fayetteville	Raleigh	Wilmington
Asheville		125	221	262	247	327
Charlotte	125		142	139	168	204
Durham	221	142		91	20	156
Fayetteville	262	139	91		65	117
Raleigh	247	168	20	65		130
Wilmington	327	204	156	117	130	

Steve drove from Raleigh to Charlotte in 3 hours. What was his average speed?

(A) 50 miles per hour
(B) 54 miles per hour
(C) 56 miles per hour
(D) 58 miles per hour 3.01

53. Patty has carefully weighed and measured the length of a licorice stick before taking the first bite and again after each bite. From her data shown in the table below, she has concluded that the weight of the remaining licorice stick is proportional to the length.

Bite Number	Length	Weight
0	304mm	28.6 grams
1	280mm	26.3 grams
2	250mm	23.5 grams
3	239mm	22.5 grams
4	202mm	? grams

After the 4th bite, the licorice stick was 202 mm long. Approximately how many grams should the licorice stick have weighed?

(A) 17 grams
(B) 18 grams
(C) 19 grams
(D) 20 grams 3.03a

54. Janice is comparing the price of three brands of olive oil. Which brand is the best buy?

Olive Oil	Size (milliliters)	Price
Brand X	709 mL	\$10.99
Brand Y	500 mL	\$8.65
Brand Z	442 mL	\$4.99

(A) Brand X is the least expensive per mL.
(B) Brand Y is the least expensive per mL.
(C) Brand Z is the least expensive per mL.
(D) Cannot be determined.

3.03b

55. If 3 out of 4 people use a certain headache medicine, how many in a city of 150,400 will use this medicine?

(A) 118,200
(B) 37,600
(C) 50,133
(D) 112,800 1.03

56. There are 5 more than twice as many boys in the weightlifting class as there are girls. There are 38 students in the class altogether. How many girls are in the class?

(A) 8
(B) 11
(C) 12
(D) 15 4.03

57. Jeff wanted to see how many of each candy flavors he had in his lunch box. He counted 25 pieces out of which he had 3 more lemon than grape, and 5 more grape than strawberry. How many strawberry candies did Jeff have?

(A) 4
(B) 9
(C) 12
(D) 15 4.03

58. Al weighs 5 pounds less than three times Little Bill's weight. Which equation represents this statement?

(A) $a - 5 = 3b$
(B) $a - 5 = b - 3$
(C) $a = 5 - 3b$
(D) $a = 3b - 5$

4.01b

59. The coordinates for the endpoints of a line segment are $(-5, 13)$ and $(-9, 21)$. Find the coordinates for the midpoint of the line segment.

(A) $(-7, 17)$
(B) $(-4, 4)$
(C) $(-2, 4)$
(D) $(4, 17)$

2.01

60. Solve: $10 + 3(2x - 6) \leq 4(2x - 7)$

(A) $x \leq -1$
(B) $x \leq -\frac{10}{7}$
(C) $x \leq -10$
(D) $x \geq 10$

4.01a

61. Factor: $4x^2 + x - 3$

(A) $(2x + 3)(2x - 1)$
(B) $(4x - 3)(x + 1)$
(C) $(2x - 3)(2x + 1)$
(D) $(4x + 3)(x - 1)$

1.01c

62. Solve the following equation: $3x^2 = 9x$

(A) $x = \{0, 1\}$
(B) $x = \{3, 1\}$
(C) $x = \{0, 3\}$
(D) $x = \{3, -3\}$

4.02

63. Solve: $4y^2 - 9y = -5$

(A) $\left\{1, \frac{5}{4}\right\}$
(B) $\left\{-\frac{3}{4}, -1\right\}$
(C) $\left\{-1, \frac{4}{5}\right\}$
(D) $\left\{\frac{5}{16}, 1\right\}$

4.02

64. Solve for y: $2y^2 + 13y + 15 = 0$

(A) $\left\{\frac{3}{2}, \frac{5}{2}\right\}$
(B) $\left\{\frac{2}{3}, \frac{2}{5}\right\}$
(C) $\left\{-5, -\frac{3}{2}\right\}$
(D) $\left\{5, -\frac{3}{2}\right\}$

4.02

65. Which ordered pair is a solution for the following system of equations?

$4x + 6y = 0$
$-2x - 5y = 4$

(A) $(6, -4)$
(B) $(-12, 4)$
(C) $(2.5, -1)$
(D) $(3, -2)$

4.03

66. Which ordered pair is not a solution for the following system of equations?

$2x - 8y = 6$
$-x + 4y = -3$

(A) $(7, 1)$
(B) $(-5, -2)$
(C) $(15, 3)$
(D) $(10, 2)$

4.03

67. The coordinates of a line segment are $(1, 6)$ and $(11, -4)$. What are the coordinates for the midpoint?

(A) $(6, 1)$
(B) $(10, 2)$
(C) $(5, 1)$
(D) $(12, 2)$

2.01

68. What is the range of the function $y = 2x - 8$ for the domain $\{10, 11, 12, 13\}$?

(A) $\{9, 9\frac{1}{2}, 10, 10\frac{1}{2}\}$
(B) $\{5, 6, 7, 8\}$
(C) $\{9, 10, 11, 12\}$
(D) $\{12, 14, 16, 18\}$

4.01a

69. For $f(x) = 3x^2 - 5x$, find $f(-3)$.

(A) 12
(B) -6
(C) 42
(D) 3

4.02

70. Yon works for Zap Electric as an electrician. When he is sent out on a job, he is told to charge a \$120 travel fee + \$80/hour. Yon uses the formula $c = 80h + 120$. Yon worked 4 hours on a job. How much did he charge?

(A) \$200
(B) \$204
(C) \$400
(D) \$440

4.01b

71. Multiply and simplify: $(x - 2)(x^2 + 3x - 9)$

(A) $x^3 + 3x^2 - 9x - 2$
(B) $4x^2 + 18$
(C) $2x^2 - 15x + 18$
(D) $x^3 + x^2 - 15x + 18$

1.01c

72. Olivia works for an air sanitizing company selling their products at a home improvement store. She makes \$12 an hour plus \$20 for every product she sells. She works forty hours a week. If she were to write a function expressing the amount of pay she receives from her place of employment each week, what would the independent variable be?

(A) the number of products she sells
(B) the amount of money she makes
(C) the forty hours a week she works
(D) Olivia

4.01b

73. Thomas ate one box of crackers plus three more crackers for a total of 225 calories. Ricardo ate two boxes of crackers plus one more cracker for a total of 325 calories. Which system of equations represents the situation above?

(A) $b + 3c = 225$; $2b + c = 325$
(B) $b = 3c + 225$; $c = 2b + 325$
(C) $b + 3 = 225$; $2c + 1 = 325$
(D) $b + 225 = 3c$; $c + 325 = 2b$

4.03

74. The graph of $y = x^2 + 2$ is changed to $y = x^2 - 2$. What happened to the graph?

(A) It moved up four spaces along the y-axis.
(B) It moved down two spaces along the y-axis.
(C) It moved down four spaces along the y-axis.
(D) It became wider.

4.02

75. $\begin{bmatrix} 0 & -2 \\ 4 & 5 \end{bmatrix} + \begin{bmatrix} -2 & -1 \\ 6 & 14 \end{bmatrix} =$

(A) $\begin{bmatrix} -2 & 1 \\ 2 & 9 \end{bmatrix}$

(B) $\begin{bmatrix} -2 & 2 \\ 24 & 5 \end{bmatrix}$

(C) $\begin{bmatrix} -2 & -3 \\ 10 & 19 \end{bmatrix}$

(D) $\begin{bmatrix} 2 & 3 \\ 10 & 19 \end{bmatrix}$

3.02

76. $\begin{bmatrix} 14 & 5 \\ -1 & 4 \end{bmatrix} - \begin{bmatrix} -2 & 7 \\ 3 & 3 \end{bmatrix} =$

(A) $\begin{bmatrix} 12 & -2 \\ -4 & 1 \end{bmatrix}$

(B) $\begin{bmatrix} 12 & 12 \\ 2 & 7 \end{bmatrix}$

(C) $\begin{bmatrix} 16 & 2 \\ 4 & 1 \end{bmatrix}$

(D) $\begin{bmatrix} 16 & -2 \\ -4 & 1 \end{bmatrix}$

3.02

77. Jack fetched seven pails, p, and six buckets, b, of water totaling 32 gallons. Jill fetched four pails, p, and nine buckets, b, totaling 35 gallons. Which system of equations represents the situation above?

(A) $7p + 6b = 32$; $4p + 9b = 35$
(B) $7p + 4b = 32$; $6p + 9b = 35$
(C) $11p + 15b = 67$; $p + b = 26$
(D) $11p + 15b = 67$; $p + b = 2$

4.03

78. The value of a computer depreciates over time. Yesterday, Sam bought a computer for \$1500, and he estimates that it will depreciate 40% over the next year. The function $f(x) = 1500(0.60)^x$ describes the value of the computer over x years. Using the function, what is the value of the computer in four years?

(A) −\$100
(B) \$194.40
(C) \$900
(D) \$5762.40

4.04

For questions 79 and 80, use the formula for growth of money in a savings account, $A = P(1+r)^t$, where A is the value of the account after t years, P is the original amount of money in the account, and r is the annual interest rate.

79. Stephany opened a savings account and put a beginning balance of \$1100 into the account. Her interest rate is 4%. About how much money will be in the account in 5 years?

(A) \$1338
(B) \$904
(C) \$3,437,500
(D) \$238

4.04

80. Franco put \$2000 in a savings account 10 years ago. He checked his balance today and noticed that he had made \$1600 in interest. Based on this information, what is the approximate interest rate on Franco's savings account?

(A) 3%
(B) 4%
(C) 5%
(D) 6%

4.04

Index